Springer Proceedings in Mathematics & Statistics

Volume 326

Springer Proceedings in Mathematics & Statistics

This book series features volumes composed of selected contributions from workshops and conferences in all areas of current research in mathematics and statistics, including operation research and optimization. In addition to an overall evaluation of the interest, scientific quality, and timeliness of each proposal at the hands of the publisher, individual contributions are all refereed to the high quality standards of leading journals in the field. Thus, this series provides the research community with well-edited, authoritative reports on developments in the most exciting areas of mathematical and statistical research today.

More information about this series at http://www.springer.com/series/10533

Vera Viana · Vítor Murtinho · João Pedro Xavier
Editors

Thinking, Drawing, Modelling

GEOMETRIAS 2017, Coimbra, Portugal, June 16–18

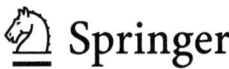 Springer

Editors
Vera Viana
Faculdade de Arquitectura and Centro de
Estudos de Arquitectura e Urbanismo
Universidade do Porto
Porto, Portugal

Associação dos Professores
de Geometria e de Desenho
Aproged
Porto, Portugal

João Pedro Xavier
Faculdade de Arquitectura and Centro de
Estudos de Arquitectura e Urbanismo
Universidade do Porto
Porto, Portugal

Vítor Murtinho
Departamento de Arquitectura
and Centro de Estudos Sociais
Universidade de Coimbra
Coimbra, Portugal

ISSN 2194-1009   ISSN 2194-1017   (electronic)
Springer Proceedings in Mathematics & Statistics
ISBN 978-3-030-46803-3   ISBN 978-3-030-46804-0   (eBook)
https://doi.org/10.1007/978-3-030-46804-0

Mathematics Subject Classification (2010): M34000, M21006, M13003, K0000X, I23036

This Springer imprint is published by the registered company Springer Nature Switzerland AG
The registered company address is: Gewerbestrasse 11, 6330 Cham, Switzerland

Preface

The International Conference Geometrias'17, held in the Department of Architecture of the University of Coimbra, between the 16 and the 18 of June, 2017, can be regarded as (another) corollary and proof that geometry stands, to this day, as a subject of the utmost importance, through which scholars, researchers, specialists and students are continuously challenged and motivated in their professional procedures, teaching practices and scientific investigations. The prominence of digital technologies in every practice related to architecture, arts and engineering is an undeniable and welcomed fact, but one may recognize that there has never been, as much as today, such an awareness on the need for a well-informed reasoning on the representational procedures as an essential requirement to ensure the conscious developments in scientific and technological researches. Conceived as one more contribution to this discussion, the leitmotif *Thinking, Drawing, Modelling* for an International Conference with a call for contributions, revealed itself as a successful strategy to bring together many scholars and investigators that actively work upon these matters and hold geometry, in its broader sense, as common concern.

The conference was a firm testimony of the importance of form-finding traditional and innovative methodologies, as well as a moment for discussions on the procedures involved in the conceptualization of objects as creative outcomes of new materialities and artistic concepts. In fact, much of the production in architecture, arts or engineering is anchored in technologies that firmly entwine with the science of representation. It is precisely within this innovative milieu that new dynamics are being generated every day, with inspiring ground-breaking ideas to stimulate more challenges, new synergies, different frameworks and inventive forms. Challenged by these new energies, the impact that virtual environments outline not only in project methodologies, but also in its concretization in space should not to be undermined.

Geometrias'17 gathered keynote speakers and authors that have been producing some of the best scientific practices concerning geometry, drawing and digital knowledge, and this, by itself, was a notable starting point that settled the pace for the quality of the contributions presented in this unique event. This book combines a selection of the outcomes of this International Conference, including three papers authored by keynote speakers and nine others, authored by scholars, researchers and students from six European countries. The following paragraphs will try to briefly summarize the content of each research.

Maurizio Barberio, in "Prototyping Stereotomic Assemblies: Stone Polysphere", demonstrates the immense potential of digital fabrication for the exploration of stereotomic architecture. Barberio illustrates the geometrical transformations through which the stone installation *PolySphere* was conceived with the spherical icosidodecahedron as point of departure, and three-dimensional modelling and algorithmic software as recurring tools.

In "Geometry and Digital Technologies in the Architecture of Herzog & De Meuron. The Project for the Stamford Bridge Stadium in London", Alexandra Castro proposes an interesting perspective on the work of these noteworthy swiss architects, identifying how the exploration of digital technologies developed in their working procedures to the current status, that significantly recognize how modern technologies bring advantages for their architectural project methodologies.

In "The Dome as Minimal Housing Unit: "Ghibli" and "D-Home" Prototypes", Micaela Colella combines the potential of three-dimensional modelling with digital fabrication towards the development of prototypes for housing units, conceived from the discretization of domes, which, as shelter modules, may undergo severe climacteric conditions.

With a long and fruitful activity in research and pedagogy concerning drawing and representation, Lino Cabezas Gelabert addresses the procedures and methods of representation explored since Middle Ages, with the lecture "Geometry and Art". The fact that, in medieval methods of representation, geometry, in its instrumental component, was recurrently explored as much as in its conceptual counterpart, might be justified by the strong tendency to mysticism of this period of human-kind's history. In fact, the methods of representation of religious architecture in that era, with many geometrized forms and rigorous metrical systems, developed into its known form, because of the scientific procedures chosen for the representation and control of space.

Soraya M. Genin's lecture, "The Vaults of Arronches Nossa Senhora da Assunção and Misericordia Churches. Geometric and Constructive Comparison with the Nave and Refectory Vaults of Jerónimos Monastery", describes hypotheses for the complex systems that might have been used to constructively resolve the enlargement of space in certain religious buildings of the first half of the sixteenth century. Regarding probable methods of drawing for the concretization of different kinds of vaults, Genin establishes interrelations between constructive methodolo-gies for arched structures and ribbed solutions. Illustrating an historical journey with a number of examples with different degrees of complexity, the author theo-rizes about the methods of drawing known and the geometry that, in its essence, allowed its materialization in space.

In "Perspective Transformations for Architectural Design", Cornelie Leopold addresses the relations between space and image from the logic of perception of the architectonic conception procedures. This paper establishes numerous connections between architecture and mathematics, examining situations where perspective drawing precedes and mediates creative workflows in which space is created and manipulated.

Joana Maia and Vitor Murtinho expose some of the the procedures adopted by a Portuguese architect, in "Ordered Creativity: The Sense of Proportion in João Álvaro Rocha's Architecture". Starting with the analysis of his creative process and working methodologies, this paper provides an understanding of the peculiarity of his working process, in which grids, as well as metric and proportional systems are intentional recurring resources with a significant impact in the architect's creativity.

Andrés Martín-Pastor and Alicia López-Martínez present the paper "Developable Ruled Surfaces from a Cylindrical Helix and their Applications as Architectural Surfaces", to demonstrate how a set of surfaces, with certain curvature and geometrical definition, are extremely appropriate to perform as light architectural structures.

Hannah Müller, Christoph Nething, Anja Schalk, Daria Kovaleva, Olivier Gericke and Werner Sobek, develop the theme "Porous Spatial Concrete Structures Generated Using Frozen Sand Formwork", a research still at an experimental stage, that focuses on the exploration of hydroplotting to conceive complex concrete forms as spatial structures or architectural objects.

José Pedro Sousa, a renowned specialist in generative geometries, through "Calculated Geometries. Experiments in Architectural Education and Research", depicts a series of examples in which geometry and technology were crucial for the development of form. Combining traditional methodologies with robotic fabrication, his research projects highlight the great potential of these innovative methods in the act of thinking and materializing architecture. For educational and research purposes, these *Calculated Geometries* emerge as powerful instruments for the exploration of new forms in architecture and reveal themselves essential to expand, into unforeseen outcomes, the magnificent art of manipulating space.

Monika Sroka-Bizoń, in "How to Construct the Red Sea?", reflects upon the concepts that led to the materialization of a successfully accomplished example of the structural importance of geometry in free-form architecture, the Museum of the History of Polish Jews, in Warsaw. The author presents the methodology explored by the Finish architects Rainer Mahlamäki and Ilmari Lahdelma who, through architectural effects of intrinsic complexity, symbolically materialize *Yum Suf,* the parting of the *Red Sea* that led to the escape of the Jews from Egypt, combining a steady regular exterior shape with interior curved surfaces, structurally intertwined with Bezier, B-spline and NURBS surfaces.

With "How to Improve the Education of Engineers—Visualization of String Construction Bridges", Jolanta Tofil presents some examples on how the exploration of CAD software and its visualization features may assist engineering students to combine their knowledge and practice for conceiving string construction bridges, with the input of architectural design procedures. In reality, the inherent potential of the possibility of previewing any structural and architectonic solution unequivocally allows us to predict better solutions, thanks to a more complete and well-informed understanding of the project, which, in itself, fosters the enhancement of the knowledge of both structure and form.

In conclusion, as all the papers combined in this book aimed to demonstrate, *Thinking, Drawing and Modelling* remain still as fundamental activities that stand in the origin of every act leading to the creation of form and its materialization. As such, geometry, drawing and the sciences of representation prevail as pertinent matters whose discussion has a great future in sight. Three-dimensional modelling and algorithmic software are magnificent proposals that are paving the way to great challenges in our world, but one cannot ignore that stereotomy, drawing, the science of representation and geometric literacy remain still as limitless inspiring sources of knowledge, so fundamental as much as inevitable, for every professional and scholar committed to the investigation and innovation in geometry-related settings.

The Editors

Porto, Portugal   Vera Viana
                                                                                      vlopes@arq.up.pt
Coimbra, Portugal   Vítor Murtinho
                                                                                       vmurtinho@uc.pt
Porto, Portugal   João Pedro Xavier
                                                                                      jxavier@arq.up.pt

Acknowledgements

The Editors, Vera Viana, Vítor Murtinho and João Pedro Xavier, express their immense gratitude for the contribution of the reviewers listed below which, through their acute experience and knowledge, certified the scientific quality, not only of the Geometrias'17 Conference, but also of its published outcomes: the book "Thinking, Drawing, Modelling" and Aproged's Bulletin #34.[1]

Giuseppe Amoruso, Dipartimento di Design—Politecnico Milano, Italy
Antonio Luis Ampliato, Universidad de Sevilla, Spain
Javier Barrallo, University of the Basque Country, Spain
Maria Francisca Blanco, Universidad de Valladolid, Spain
Vasco Cardoso, Faculdade de Belas Artes da Universidade do Porto, Portugal
Graciela Colagreco, Universidad Nacional de la Plata, Argentina
Manuel Couceiro da Costa, Faculdade de Arquitectura da Universidade de Lisboa, Portugal
Giuseppe Fallacara, Politecnico di Bari, Italy
Lino Cabezas Gelabert, Facultat de Belles Arts, Universitat de Barcelona, Spain
Soraya Genin, ISCTE-Instituto Universitário de Lisboa, Portugal
Filipe Gonzalez, Escola de Arquitectura e Artes da Universidade Lusíada de Lisboa, Portugal
Encarnación Reyes Iglesias, Universidad de Valladolid, Spain
Andreas Kretzer, Technical University of Kaiserslautern, Germany
Cornelie Leopold, Technical University of Kaiserslautern, Germany
João Ventura Lopes, ISCTE—Instituto Universitário de Lisboa
Pedro Martins, Faculdade de Arquitectura da Universidade do Porto, Portugal
Luís Mateus, Faculdade de Arquitectura da Universidade de Lisboa, Portugal
Helena Mena Matos, Faculdade de Ciências da Universidade do Porto, Portugal
Riccardo Migliari, La Sapienza, Università di Roma, Italy

---

[1]The digital version of this publication is available in https://www.aproged.pt/geometrias17/boletim34digital.pdf.

Gonçalo Canto Moniz, Faculdade de Ciências e Tecnologia da Universidade de Coimbra, Portugal

Vítor Murtinho, Faculdade de Ciências e Tecnologia da Universidade de Coimbra, Portugal

Ratko Obradovic, University of Novi Sad, Serbia

Alexandra Paio, ISCTE—Instituto Universitário de Lisboa Portugal

Teresa Pais, Faculdade de Ciências e Tecnologia da Universidade de Coimbra, Portugal

Eliana Manuel Pinho, CEAU—Faculdade de Arquitectura da Universidade do Porto, Portugal

Luís Marques Pinto, Universidade Lusíada do Porto, Portugal

Francisco González Quintial, Escola Técnica Superior de Arquitectura Universidad del País Vasco, Spain

Rinus Roelofs, Netherlands

José Ignacio Rojas Sola, University of Jaen, Spain

José Pedro Sousa, Faculdade de Arquitectura da Universidade do Porto, Portugal

Vera de Spinadel, Mathematics and Design Association's, Argentina

Hellmuth Stachel, Technische Universität Wien, Austria

António Oriol Trindade, Faculdade de Belas Artes da Universidade de Lisboa, Portugal

Pedro de Azambuja Varela, Faculdade de Arquitectura da Universidade do Porto, Portugal

Gunter Weiss, Technische Universitat Dresden, Germany

João Pedro Xavier, Faculdade de Arquitectura da Universidade do Porto, Portugal

Contents

Prototyping Stereotomic Assemblies: Stone Polysphere 1
Maurizio Barberio

**Geometry and Digital Technologies in the Architecture of Herzog & de
Meuron. The Project for the Stamford Bridge Stadium in London** 13
Alexandra Castro

**The Dome as Minimal Housing Unit: "Ghibli" and "D-Home"
Prototypes** 29
Micaela Colella

Geometry and Art 41
Lino Cabezas Gelabert

**The vaults of Arronches Nossa Senhora da Assunção and Misericórdia
churches. Geometric and constructive comparison with the nave
and refectory vaults of Jerónimos Monastery** 61
Soraya M. Genin

Perspective Transformations for Architectural Design 77
Cornelie Leopold

**Ordered Creativity: The Sense of Proportion in João Álvaro Rocha'S
Architecture** 91
Joana Maia and Vítor Murtinho

**Developable Ruled Surfaces from a Cylindrical Helix
and Their Applications as Architectural Surfaces** 107
Andrés Martín-Pastor and Alicia López-Martínez

**Porous Spatial Concrete Structures Generated Using Frozen
Sand Formwork** 121
Hannah Müller, Christoph Nething, Anja Schalk, Daria Kovaleva,
Oliver Gericke, and Werner Sobek

Calculated Geometries. Experiments in Architectural Education and Research 131
José Pedro Sousa

How to Construct the Red Sea? 145
Monika Sroka-Bizoń

How to Improve the Education of Engineers—Visualization of String Construction Bridges 155
Jolanta Tofil

Prototyping Stereotomic Assemblies: Stone Polysphere

Maurizio Barberio

Abstract The paper describes the parametric design, fabrication and construction of the Stone PolySphere, an installation that investigates the potential of digital fabrication applied to the stone industry. It is a lithic sphere with a diameter of 1.4 m, composed by a massive hemisphere below and a stereotomic hemisphere above. The prototype ideally summarizes, in a single object, the two big trends in stone architecture: the megaliths (below) and the stereotomic assemblies (above). The upper part is a micro-architecture that represents a domed space. The research also shows a new workflow for the realisation of non-reciprocal stereotomic geodesic assemblies, in which are used complex holed voussoirs generated by means of *simple* morphing operations and recursive subdivisions. The fabrication of the voussoirs consists in milling of several layers of stones glued together. The results are critically discussed, and the implications of the parametric design process are pointed out.

Keywords Stereotomy · Stone · Parametric design · Digital fabrication · Sphere

1 Background

Throughout history, the domed space has always been the ideal field for more sophisticated and complex studies about the construction of architecture. This is particularly true for the stereotomic architecture. Stereotomy, from the Greek: $\sigma\tau\varepsilon\rho\varepsilon\acute{o}\varsigma$ (*stereós*) "solid" and $\tau o\mu\acute{\eta}$ (*tomē*) "cut", is the art and science of cutting three-dimensional solids into particular shapes [1]. The intrinsic quality of the domed architecture resides in its immediate ability to define measurable areas, which can serve as the endpoint for the indeterminacy of a generic space outside the building [2]. Historically, a stone dome is usually built laying voussoirs by rows. This building method is very common throughout the history of architecture and construction, and it has been widely used since ancient times [3]. More recently, several scholars have studied another way to build a stone dome, by using geodetic tessellations. Historically, they

M. Barberio (✉)
New Fundamentals Research Group, Politecnico di Bari, Bari, Italy
e-mail: mb@newfundamentals.it

© Springer Nature Switzerland AG 2020
V. Viana et al. (eds.), *Thinking, Drawing, Modelling*,
Springer Proceedings in Mathematics & Statistics 326,
https://doi.org/10.1007/978-3-030-46804-0_1

have been used in architecture since the last century, mostly thanks to the work of Buckminster Fuller. Regarding stereotomy, it is possible to use two types of geodesic tessellations: reciprocal and non-reciprocal. Reciprocal tessellations employ interlocking blocks with a specific geometry that are arranged such that the inner end of each block rests upon and is supported by the adjacent block. On the other hand, non-reciprocal tessellations ensure stability thanks to the friction between the contact faces of blocks and the overall shape of the structure itself. The first built example of a geodesic stereotomic dome made of interlocking reciprocal stone blocks is the *Bin Jassin Dome*, built in 2012 in Qatar, and designed by Giuseppe Fallacara [4]. An example of a geodesic non-reciprocal dome has been recently studied by Roberta Gadaleta [5] for her doctoral dissertation.

2   Research Topic

Stone PolySphere is an installation that investigates the potential of digital fabrication applied to the stone industry. It is a lithic sphere with a diameter of 1.4 m, composed by a massive hemisphere below and a stereotomic hemisphere above. The prototype ideally summarizes, in a single object, the two big trends in stone architecture: the megaliths (below) and the stereotomic assemblies (above). The upper part of the PolySphere is a micro-architecture that represents the domed space.[1] As such, this case-study is useful to investigate the following themes:

- Using not reciprocal geodesic tessellations to build domed spaces;
- Producing voussoirs through milling of several layers of stones glued together;
- Generating complex holed voussoirs through simple morphing operations and recursive subdivisions;
- Setting specific algorithms in order to automatically generate the files necessary for the fabrication by means of CNC machines.

3   Computational Workflow

The computational workflow is based on a parametric code which has a base polyhedron as input, and all the three-dimensional lithic elements to be fabricated as output. The software used to accomplish the algorithm is Grasshopper™, a parametric plug-in developed for the commercial software Rhinoceros™. From a geometric point of view, Stone PolySphere (Fig. 1) is generated by the geodetic projection of a polyhedron, the icosidodecahedron, whose faces are tessellated as follows:

---

[1]It is important to specify that the research intent was not to make a scaled-down model of a real dome, but a sphere divided into voussoirs like a real dome. For this reason, structural analysis and material tests were not carried out.

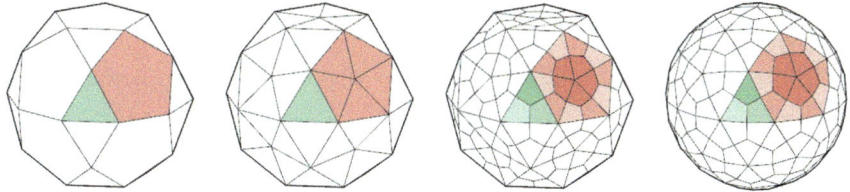

Fig. 1 Subdivision steps and geodesic projection of the base polyhedron

All the pentagonal faces were divided into triangles;

1. Each of these triangles, as well as the icosidodecahedron's triangular faces, were divided into three kites, by joining the middle of each side to the centroid of the triangle itself (Catmull–Clark subdivision, level 1), thus generating the quadrilateral meshes;
2. All vertices of the kite-shaped meshes are projected onto the surface of the circumsphere. These vertices moved along the direction of the vectors formed by connecting each vertex with the centroid of the sphere.

This geometric construction is identical for the whole sphere, but is processed in two different ways for the lower part (massive) and the upper one (stereotomic). From a computational point of view, however, the parametric definition necessary to the creation of the three-dimensional model is conceptually identical. In fact, both the voussoirs of the upper part and the base-surfaces are obtained through the following operations: recursive subdivision of the base-pattern and morphing of the base-pattern used to "populate" the surface. Morphing operations are commonly used in parametric modelling, and they are generally employed to tessellate a given surface, dividing it into portions of square or rectangular-based prisms (cuboids or "boxes"). To each box, a generic geometric base-pattern is associated (a sphere, a cube, etc.). In other words, this operation is used to transfer a geometric base-pattern—usually three-dimensional—onto a tessellated surface. The process is well known among Grasshopper's users [6]. Figure 2 shows the typical workflow used to tessellate a surface through morphing operations.

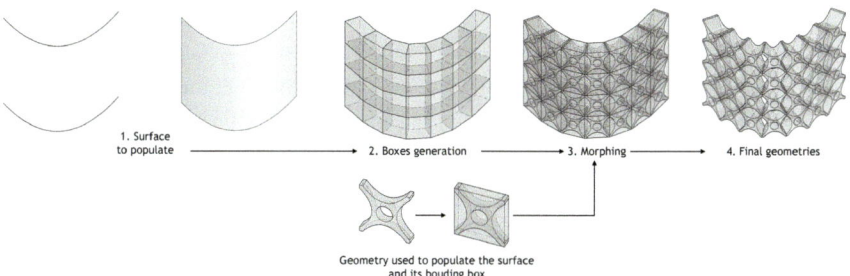

Fig. 2 Workflow for tessellating a generic surface by means of morphing operations

Most of the time, however, the process is used more on formal speculations than in constructive purposes. This happens because it is possible to tessellate the base-surface using only quadrangular-based prisms and the tessellation of the surface is not generally optimized by the geometric and constructive point of view. Consequently, this process of modelling leads to non-optimized tessellations of free-form surfaces, obtaining structural patterns deformed by morphing operations that not only are inaccurate from a formal point of view, but also not easily manufacturable.

However, this research aims to demonstrate that this *simple* modelling strategy can lead to interesting and accurate results, able to associate morphing operations with more complex spatial tessellations, as the kite-based prismatic explored in this research. As such, the workflow outlined for generating the voussoirs of the stone PolySphere was the following (Fig. 3):

1. The input mesh was analysed to check the flatness of the faces. In order to optimize the three-dimensional model for fabrication purposes by means of CNC machines, it is recommended that each voussoir has, at least, its inner face flat. This is fundamental to put the raw block on the CNC base platform and process it without rotations or the aid of supports;
2. If the intradosal faces are not flat, planarity is obtained if the distance between face diagonals is 0. This process was achieved by moving the points of each face until they were all in the same plan. The mesh was then processed using the add-on *Kangaroo Physics* [7];
3. Afterwards, the offset mesh was generated;
4. The boxes used for the morphing operations were generated;
5. The base voussoir, which will be subjected to morphing operations, had to be modelled in this step. In the specific case of the stone PolySphere, the basic voussoir was modelled starting from a squared mesh divided into four triangles, with two of them perforated. The basic voussoir was intentionally modelled with the minimum number of faces possible, in order to be gradual and recursively divided through an algorithm based on the Catmull-Clark algorithm [8]. This allowed us to obtain a uniformly smoothness for the voussoirs at the point where they are perforated, increasing the number of subdivisions of the coarse mesh.
6. Generation of the voussoirs to be produced.

It is important to point out that the use of this parametric workflow is justified when: the geometry (the voussoir) to be transferred to the tessellated surface is topologically equivalent to all the blocks or families of voussoirs; there is no need to locally manage the variation of one of the fundamental parameters of the geometry to be transferred (for example, in this case, the size of the openings); and the geometry to be transferred onto the tessellated surface has such a degree of complexity, that obtaining it with other modelling methods would be very difficult or even impossible.

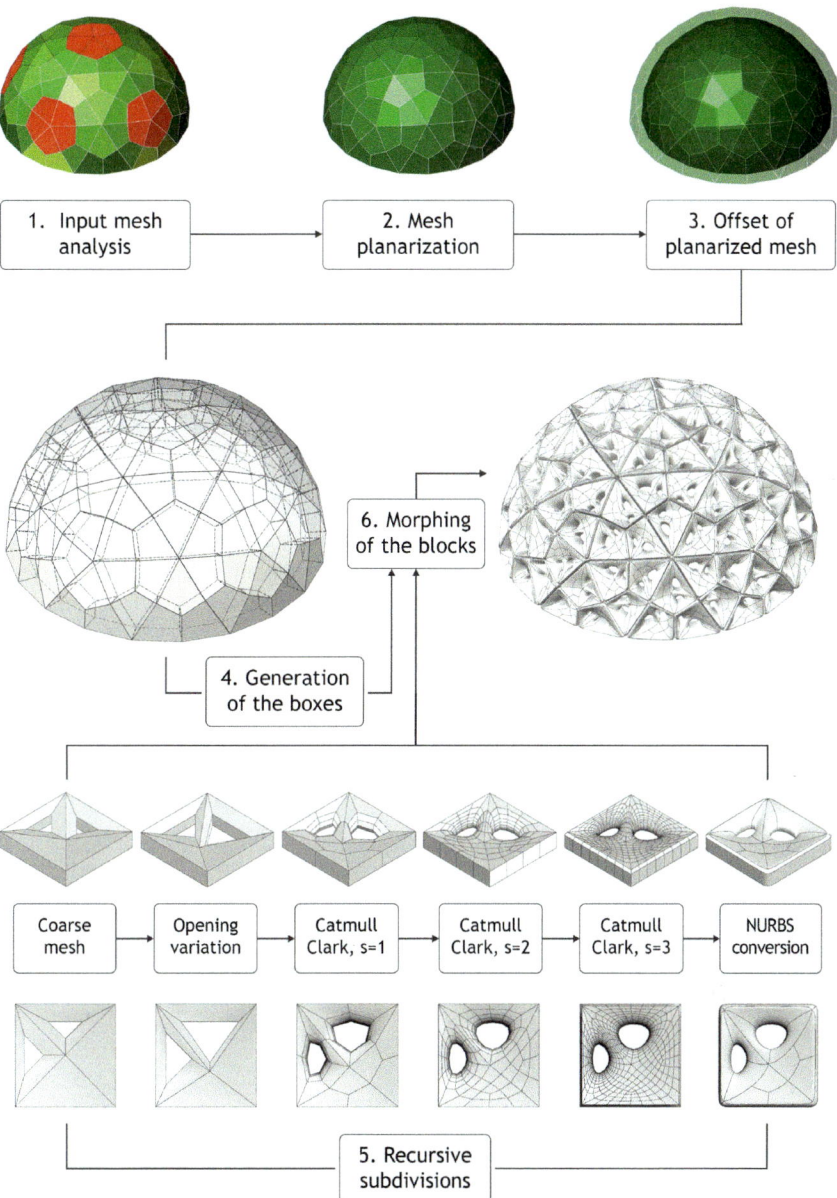

Fig. 3 Workflow for generating the blocks (voussoir) of Stone PolySphere

4 Prototypation

The massive hemisphere was obtained by sawing and milling a stone piece made of *Pierre Bleue de Savoie* (1.4 m wide and 0.8 m high); the stereotomic part consisted of 120 perforated voussoirs, fabricated by milling four blocks of stone, containing 30 voussoirs each (with the average size of 1.60 m and 10 cm thick). The CNC machine Tc 1100, produced by Thibaut SA, allowed the complex machining (milling and sawing) of all the parts. The voussoirs are constituted by two glued layers of stone, returning a changing two-colour effect depending on the morphology of the voussoir itself. The stones used were *Pierre Bleue de Savoie* for the darker part, and *Blanc d'Angola* for the clearer part (Fig. 4).

To facilitate the production of the voussoirs, the algorithm was set in order to automatically obtain the STL file of the grouped ashlars on the plane, prepared for fabrication. An additional benefit from having generated the model through a parametric workflow was the need to constantly update the geometry of the voussoirs based on the information supplied by the company involved in the construction of the prototype, Thibaut SA. A continuous exchange of feedbacks has led to regenerate the model continuously, according to the fabrication constraints. This real-time optimization process between designers and production would not have been possible without a computational approach.

CNC circular saw blade and five-axis milling have been used to fabricate both the massive and the stereotomic part. For the massive part, the fabrication steps were:

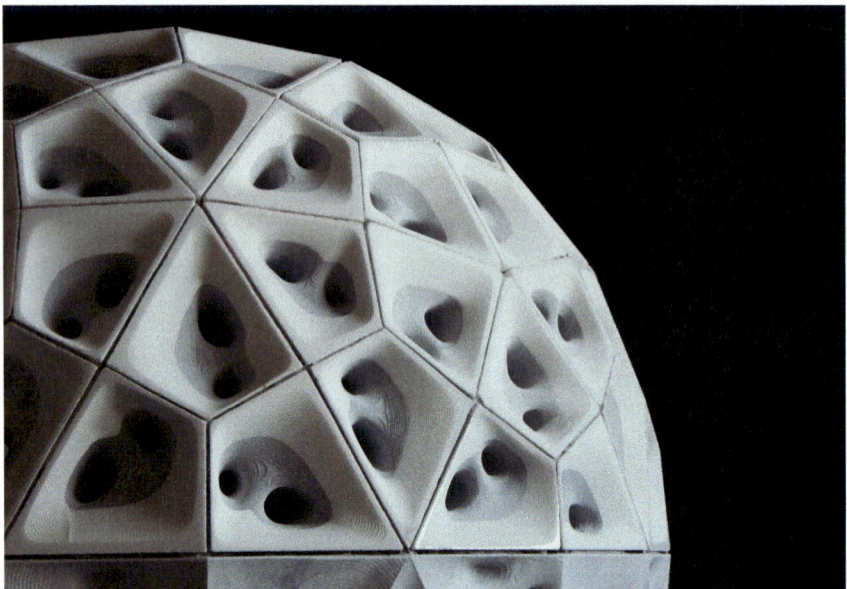

Fig. 4 Detail of the stereotomic part with the two stones used

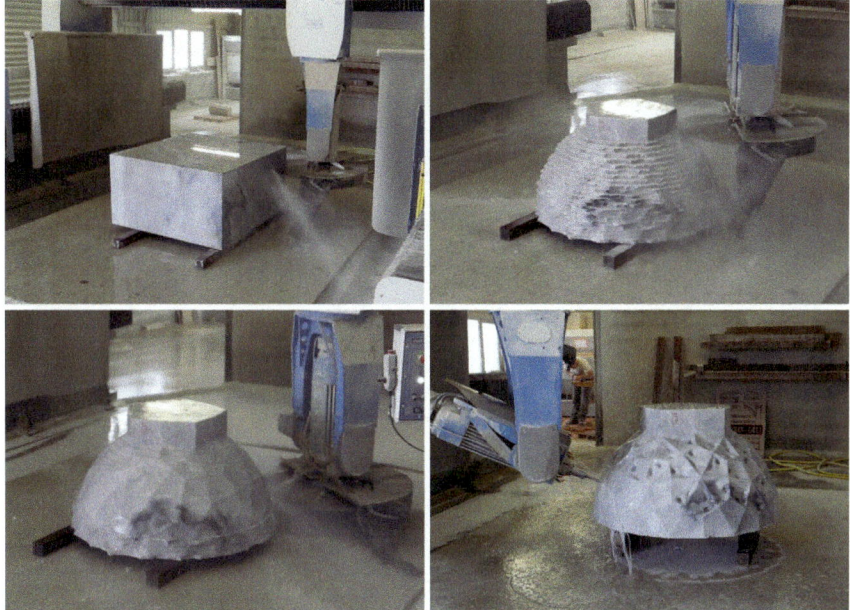

Fig. 5 Fabrication steps of the massive part

cutting of the initial parallelepiped block by means of a circular saw blade; and milling of the quadrilateral patterns constituting the sphere (Fig. 5). For the stereotomic parts, the fabrication steps were: milling of the holed parts of the voussoirs, processed in groups of 30 elements; cutting of the contact faces of the voussoirs by means of a circular saw blade (Fig. 6). Figure 7 shows a detail of the milling paths over the massive part.

The assembly was accomplished with the aid of a centring made of high-density polystyrene. The centring was designed in order to be removed before placing the keystone (Fig. 8). The assembly then revealed a series of unexpected problems. The construction of geodesic stereotomic domes is more challenging, because the voussoirs are unstable during the process. Furthermore, the tolerance between voussoirs has caused an increment of errors during the assembly, instead of being a way to reduce the unavoidable fabrication inaccuracies.

5   Conclusions

Concluding, the computational approach adopted has allowed:

- The possibility to modify the geometry of the voussoir (or the tessellation of the sphere) and regenerate the model in real time;

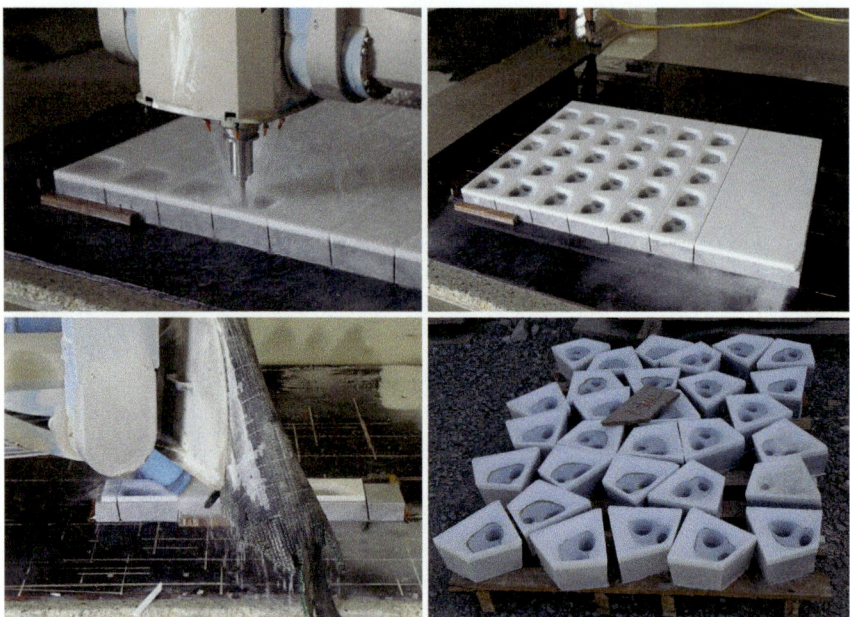

Fig. 6 Fabrication steps of the stereotomic part

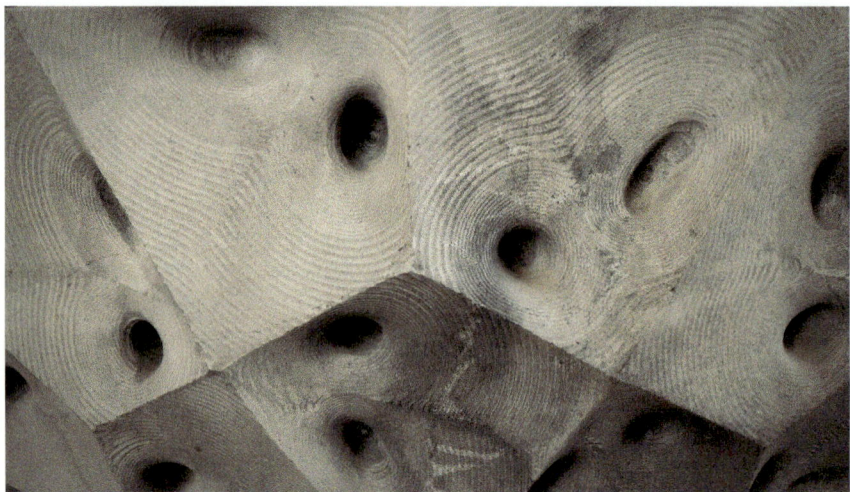

Fig. 7 Milling paths over the massive part

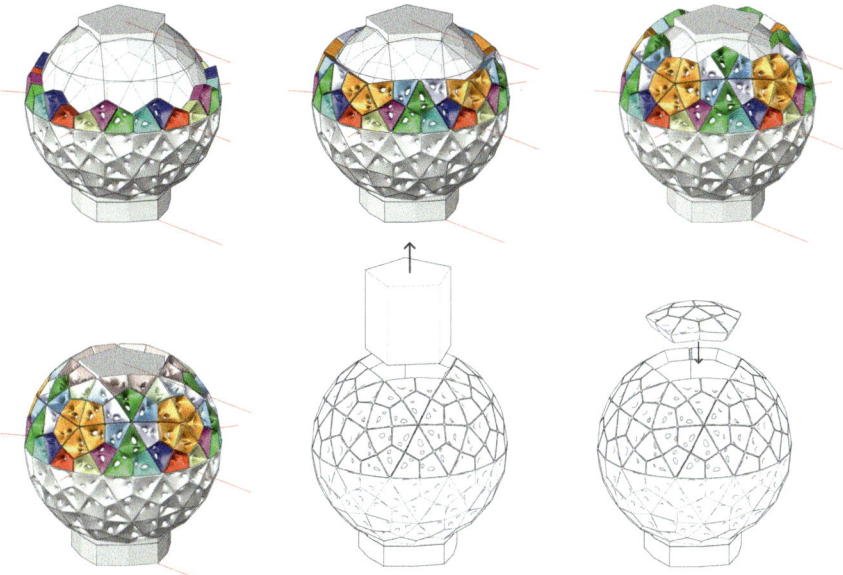

Fig. 8 Assembly steps of the stereotomic part. The colours indicate the different voussoirs types

- The possibility to get an automatic generation of rows of voussoirs grouped on the floor, ready to be made with numerical control, regardless of the geometry of the voussoir to be applied on the tessellated sphere;
- The regeneration of production files, optimized on the basis of the feedback provided remotely by the various actors involved in the prototype realization, avoided time-consuming problems and the slowdown of the production schedule.

Contextually, and for completeness of the discussion, it is necessary to note that, with the software used (Rhino 5.0 and Grasshopper 0.9), it is possible to tessellate a generic surface using boxes constituted by more than 8 vertices using exclusively the "user objects" developed by David Mans [9]. A possible progress of research about the gluing process presented in this paper might be to use several layers of stones and different thicknesses, in order to obtain an even greater lithic variability. This aspect suggests also to consider if it is possible to stick different wasted slabs and use them as new blocks, in order to combine new aesthetic possibilities and sustainability (Fig. 9). Other possibilities to be investigated are related to the connection between the size of the openings of blocks and the level of natural illumination of the interior spaces. Lastly, it is important to underline that it is possible to use the parametric strategies explained in this paper in order to accomplish complex vaulted or domed spaces, using both additive and subtractive fabrication technique (Fig. 10).

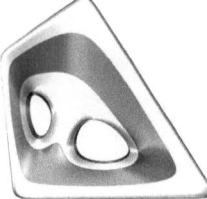 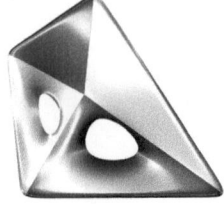 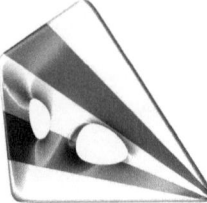

Fig. 9 Examples of different chromatic combinations (light/dark)

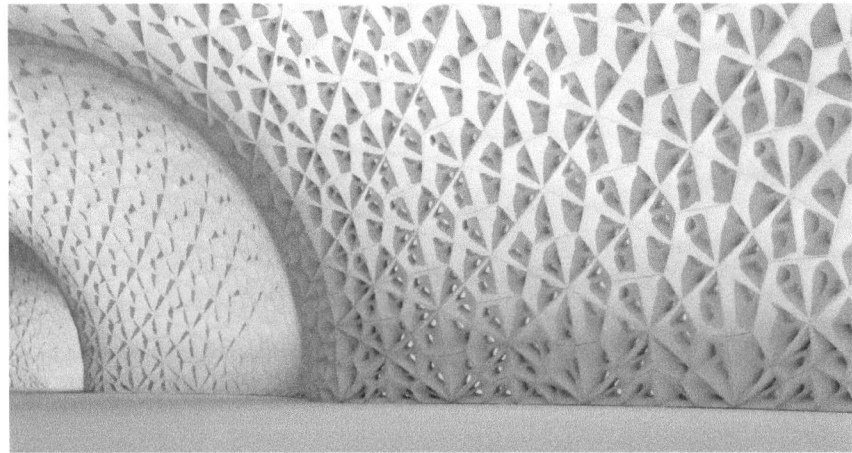

Fig. 10 Example of complex vaulted space, generated with the described method

Acknowledgements The project was developed under the scientific supervision of Prof. Giuseppe Fallacara and it is part of Barberio's doctoral dissertation" *Nuove Frontiere dell'Architettura in Pietra*" (New Frontiers of Stone Architecture). Barberio and Fallacara are co-designers of the Stone PolySphere. The author would like to thank: Yelmini for the stone supplying; Chevrin-Geli for hosting the construction of the sphere; Thibaut SA represented by Jacques Thibaut (Chairman), Alain Calas (Technical Director), Bruno Combernoux (Service Engineer) and Claire Capel (Coordinator). Finally, the author would like to thank Prof. Alberto Pugnale as co-supervisor of his doctoral thesis.

References

1. Fallacara, G.: Verso una progettazione stereotomica. Nozioni di Stereotomia, Stereotomia digitale e trasformazioni topologiche: ragionamenti intorno alla costruzione della forma. Aracne Editrice, Rome (2007)
2. Fallacara, G.: Architectural stone elements. Research, design and fabrication. Presses des Ponts, Paris (2016)

3. Galletti, S.: Stereotomy and the Mediterranean: Notes Toward an Architectural History. Mediterr. Int. J. Transf. Knowl. **2**, 73–120 (2017)
4. Fallacara, G.: Stereotomy. Stone Architecture and New Research. Presses des Ponts, Paris (2012)
5. Gadaleta, R.: Study of tradition and research of innovative stereotomic bond for dome in cut stone. In: INTBAU International Annual Event, July 2017, pp. 1262–1270. Springer, Cham (2017)
6. Tedeschi, A.: AAD_Algorithms-Aided Design. Parametric Strategies using Grasshopper. Le Penseur Publisher, Brienza (2014)
7. Piker, D.: Kangaroo: form finding with computational physics. Archit. Des. **83**(2), 136–137 (2013)
8. Catmull, E., Clark, J.: Recursively generated B-spline surfaces on arbitrary topological meshes. Comput. Aided Des. **10**(6), 350–355 (1978)
9. Retrieved in August, 2017 from http://www.neoarchaic.net/mesh

Maurizio Barberio is an architect and Ph.D. in architectural design. Barberio's research is focused on the relationship between digital architecture and fabrication innovation in the stone field. He is a partner and co-founder of "New Fundamentals Research Group" as well as co-founder of the architecture firm "Barberio Colella ARC".

Geometry and Digital Technologies in the Architecture of Herzog & de Meuron. The Project for the Stamford Bridge Stadium in London

Alexandra Castro

Abstract The present paper stems from the research we are developing in our Ph.D. work. From the study of concrete examples, we are particularly interested in understanding the impact that, in the last 25 years, the technological innovations and changes in aesthetic paradigms and spatial concepts had in architecture, interfering in the way Geometry, as a support tool for design, composition and construction of architectural objects, is used in the creative act. In this sense, the present study proposes a specific reading of the work of Herzog & de Meuron (HdM). Through the analysis of the recent project for the new Stamford Bridge Stadium in London, we intend to recognize which geometry lays behind the architectural forms, in order to gauge its role in the construction of the project and understand how new technologies can become a driving force in the exploration of this design tool.

Keywords Geometry · Architecture · Computation · Herzog & de Meuron

1 Herzog & de Meuron

1.1 Project Practice

The architectural practice of HdM is based on a traditional approach that, at least since Vitruvius, considers drawing as a fundamental instrument of the project's research, but integrating, in its methodology, the most advanced digital tools. In this sense, an analysis of HdM's work allows us to decode and understand the methodological transformations that have occurred in architecture in the last years and are transversal to contemporary practices.

Despite the overgrowth of their office, over the years, and the fact that, nowadays, it is internationally recognized as one of the most prestigious, HdM consider architecture within a disciplinary scope. For the Swiss architects, architecture is understood

A. Castro (✉)
Centro de Estudos de Arquitectura e Urbanismo and Faculdade de Arquitectura da Universidade do Porto, Porto, Portugal
e-mail: macastro@arq.up.pt

© Springer Nature Switzerland AG 2020
V. Viana et al. (eds.), *Thinking, Drawing, Modelling*,
Springer Proceedings in Mathematics & Statistics 326,
https://doi.org/10.1007/978-3-030-46804-0_2

as an artistic practice with a history and a tradition, making only sense if defined and reinvented from itself. As Jacques Herzog argued "(...) we're pupils of Aldo Rossi and continue to pursue this approach (...). We always proceeded from architecture and didn't just tackle it out of an onerous sense of duty, as other well-known and innovative architects of our generation did and also proclaimed accordingly" [1].

Even though they explore the most advanced technological systems and recognize that we live in a time when everything has become less certain and less stable, HdM are not interested in project exercises of individualistic style. When questioned about the distinction, in architecture, between the specificity and difference as a distinctive system, Jacques Herzog explained that

> (...) many of our colleagues live with an imperative of existence, of being seen and therefore of producing distinctive signs. This exerts enormous pressure, and is a source of terrible fear. They are capable of creating such signs with some force but increasingly it is difference within sameness - a caricature of difference. It creates architecture that is egoistic to the point of caricature [2].

In this sense, HdM are against the generic character of much of contemporary architecture, most of the times pre-determined by an author's style and a desire of being in the technological forefront, trying to ensure that each project has a unique and distinctive identity. Understood as a specific response to a particular context, the projects of HdM start from a strong conceptual base and an attempt to capture and highlight the specificities of the site to constitute itself as part of a whole.

1.2   Project Theory

Over a period of 40 years, it is possible to identify a turning point in HdM's work.

Since 1995, when they were chosen to design in London the Tate Modern, the Swiss architects began to receive global scale charges that created a need for a readjustment of the office's structure and of the way in which each teamwork manages the development of each project.

Meanwhile, over the years, in their process of work, which has always been hybrid and eclectic, combining all the available instruments from the most primitive to the most technologically advanced and where physical models have a great preponderance, digital tools have been incorporated in an increasingly effective way.

This issue gained particular relevance when Kai Strehlke, in 2005, created the Digital Technology Group (DTG) and became responsible for integrating in the work of the office new geometric and technical methods, both for design and fabrication. DTG is an in-house team responsible for different fields such as parametric design and scripting, digital fabrication, computer-aided design management and geometric support, building information modelling, and visualization and video.

As previously mentioned, HdM projects stand out by their uniqueness and by the strong conceptual component that supports them and that, throughout the project, is progressively transformed into a viable architectural scheme. This means that the main focus of the work is always placed on architecture and that digital tools, defined specifically for each project, are developed with the main purpose of ensuring that the concept works and is coherent. More than being specialized in creating free-form surfaces, DTG aims, above all, to find the most appropriated drawing strategies for each architectural design, being able to handle with complex geometries in a flexible and adaptable way [3].

2 Stamford Bridge Stadium

The project for the Stamford Bridge stadium, started in 2014 by HdM and still under development, aims for the improvement of the existing stadium, by the expansion of the existing capacity to 60,000 spectators, and the general reconfiguration of the facilities.

The stadium is located in the west part of London, in a district that has been hosting Chelsea Football Club since its foundation in 1905. The current facilities are the result of a sum of different buildings, deriving from the successive interventions over the years. Apart from being obsolete and functionally unsuitable, the existing set has a fragmented image that mirrors the various construction times and does not favour the quality of the venue nor the prestige of the football club itself.

HdM project forecasts the demolition of the existing facilities and the construction of a unitary volume that, approaching the limit of the current stadium footprint, will host the different valencies of the sport's building (Fig. 1). The uniform language of the new stadium architecture aims to make it stand out from the urban landscape as a recognizable landmark as well as to reinforce the identity of the club in its place of origin.

As a starting point for the analysis of this project and the understanding of its geometry, we think it is important to highlight three premises related to the site that, in our opinion, were decisive on the development of the architectural solution.

Firstly, the Conservation Areas.

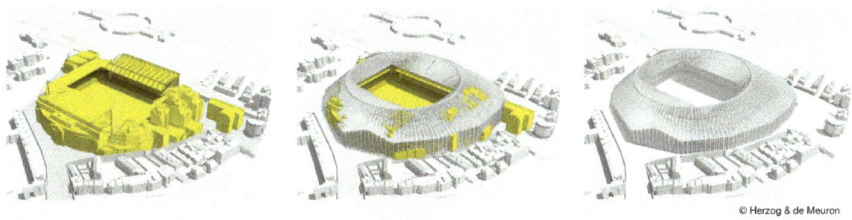

© Herzog & de Meuron

Fig. 1 Current stadium volume; overlay of current stadium and proposal; proposed volume, Herzog de Meuron (2015), *Stamford Bridge Grounds. Detailed Plan Application. Design and Access Statement* [4, p. 40]

The stadium grounds are surrounded by a considerable number of conservation areas which, due to their unique historical and architectural interest, require the preservation, valorization and guarantee of the integrity of their environment. A condition that, inevitably, interferes with the volume of the new stadium, its architectural composition and the way the building "shows" itself to the public space.

As a second point, we refer to the intervention area and its limits.

The grounds are enclosed by two railway cuttings, the metro to the north and the train to the east, which runs over the Counter's Creek, an old tidal tributary of the Thames that was canalized in the nineteenth century, and two historic boundary walls, to the south and west, which mark the limits with the adjoining residential properties. While the existing boundary walls were integrated in the project, in what refers to the railway cuttings, the designers decided to use the space above the tracks by capping the cuttings with pedestrian platforms. This operation permits to gain an exterior area surrounding the stadium that compensates the inevitable increase of the building's volume. The site's long-established configuration was determinant for the definition of the building shape, whose outline establishes a straight dialogue with the limits of the pre-existence.

The last premise refers to the pitch and the fact that, in the new project, it will maintain its historical location. This option, based on the intention of ensuring, within the arena, a continuity with the previous generations of the stadium and the recognition of the identity of the game's space by the adepts, places limitations on the orientation of the new building and, above all, on its position regarding the surrounding.

2.1 Geometry

As stated in the detailed planning application, HdM claim that the architecture of the new Stamford Bridge stadium: "will be a synthesis of the external demands of this extraordinary site with the internal requirements of a contemporary arena" [5].

In our view, this statement summarizes that which is the base geometry of the project. A rectangle and an irregular polygon synthesize the shape of the building, enclosing in its overlap the duality present in architecture.

The interior rectangle of the pitch is the reference of an orthogonal system from which is organized the design of the stands and the spectator amenities located under their structure (Fig. 2). The stadium bowl is divided in four stands, parallel to the sides of the rectangle, and the corners transition is made through the perpendiculars to the bisectors of the rectangles' angles.

On the outside, the faceted polygon that encloses the volume of the building reveals, in its irregularity, a response to the constraints of the site. The apparent randomness of this figure is actually controlled by alignments, symmetries, parallelisms and purposeful direction breaks. Regarding the relation with the surrounding contour, this geometric figure draws intentionally the outer area around the stadium, welcoming the public and driving it along circulation spaces.

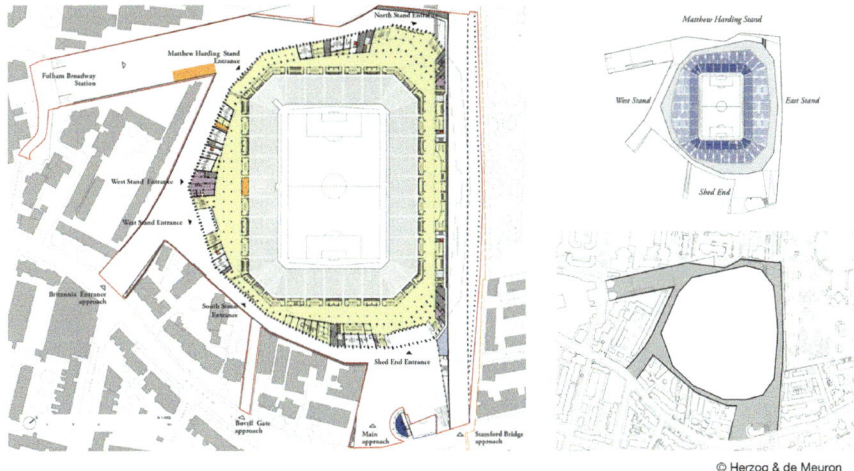

© Herzog & de Meuron

Fig. 2 Entrance plan; stadium bowl; publicly accessible space, Herzog de Meuron (2015), *Stamford Bridge Grounds. Detailed Plan Application. Design and Access Statement* [4, pp. 105, 49 and 48]

This system is overlapped by another radial system (whose origin is in the centre of the pitch) that gives body to the structure of the stadium and materializes itself on the external vertical brick piers divided in primary (90 × 213 cm) and secondary (55 × 145 cm) (Fig. 3).

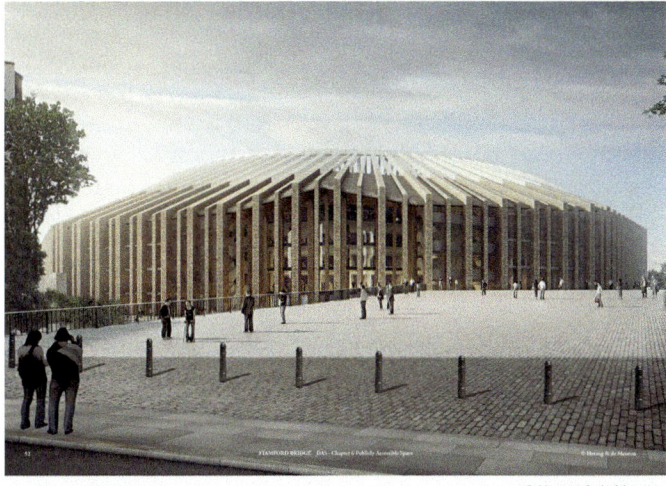

© Herzog & de Meuron

Fig. 3 Shed End Entrance, Herzog de Meuron (2015), Stamford Bridge Grounds. Detailed Plan Application. Design and Access Statement [4, p. 52]

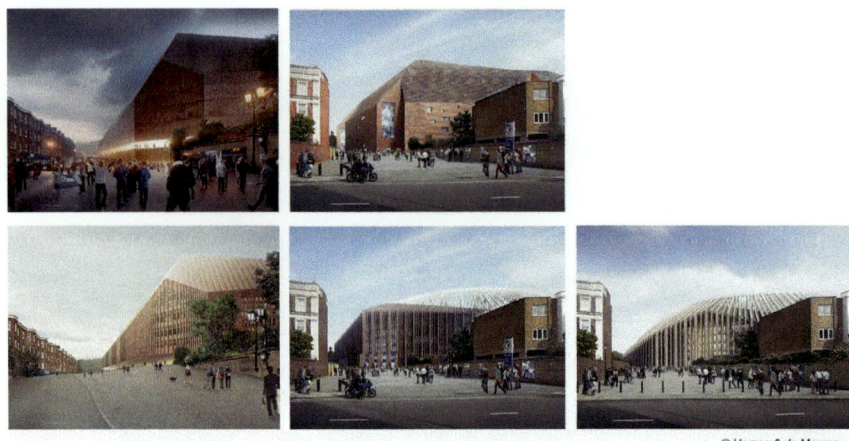

© Herzog & de Meuron

Fig. 4 April '14, June '14, July '14, September '14 and August '15 proposals, Herzog de Meuron (2015), *Stamford Bridge Grounds. Detailed Plan Application. Design and Access Statement* [4, pp. 36 and 37]

Placed alternately by size, the 264 piers, together, draw the building facade, giving it, in its abstract repetition, a unitary image capable of making the stadium a symbol of the football club and a distinctive element in the city's landscape. Simultaneously, from the point of view of the volume and its impact on the surroundings, this geometric option reveals itself to be strategic by breaking visually the scale of the building.

Observing the images regarding the project's evolution (Fig. 4), we can notice that the design of the volume, characterized in a first phase by an opaque exterior responsible for a monolithic appearance, was progressively dematerialized, in an intermediary stage by the introduction of grids associated with the openings, until an effective lightness is acquired in the final proposal of August 2015, when the geometry of a radial system took shape, becoming coincident with the building structure. The radial system of the piers, seen directly from the front, gives a transparent and light expression to the building, while, at the same time, gives it thickness and solidity when viewed in profile.

The radial system is complemented by a circular ring, located in the stadium roof structure, with the same centre, reinforcing it from the design's perspective, and complementing it by giving it sense from a structural point of view.

The facade columns that support the building rise to the roof where they articulate, through precast and brick-cladding rafters, with a set of radial trusses that advance 50 m above the bowl. This is the "double" steel ring (tension and compression) that will join all the radial trusses that materialize the inner covering of the stadium arranged over the stands.

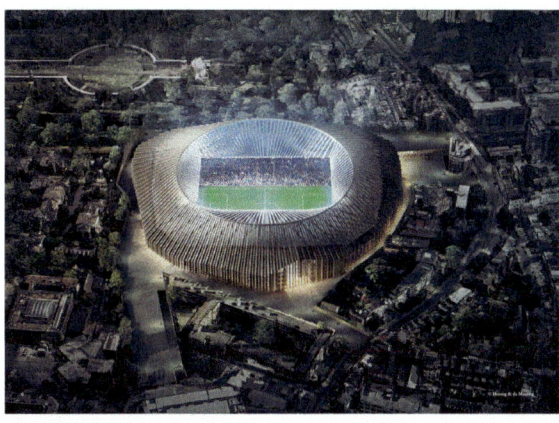
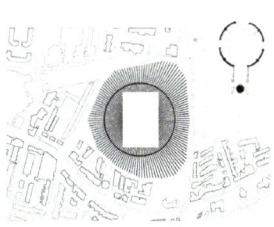

© Herzog & de Meuron

Fig. 5 General view of the stadium; circular roof ring and Brompton Cemetery arcade, Herzog de Meuron (2015), *Stamford Bridge Grounds. Detailed Plan Application. Design and Access Statement* [4, pp. 6 and 50]

Besides its structural function, from a volumetric point of view, this geometric figure, which possesses a strong centrality, has the role of unifying the whole complexity resulting from the organic and responsive character of the building's exterior (Fig. 5). This can be considered as a formal gesture of a great clarity that reveals the ability of geometry to structure the design, endowing it with an order and an aggregating reference.

Conceptually, HdM also reveal that, underlying the introduction of this element, it's the interest in establishing a dialogue with the circular geometry of the Brompton Cemetery arcade, located on the east side of the stadium.

The project is marked, in its different stages of development, by a continuous concern in reducing the scale of the volume, in order to minimize its impact on the neighbouring buildings. This aspect is achieved not only by the strategic decision in "sinking down" the stadium, lowering the pitch 4 m, but also by the permanent sculpting of the general shape of the volume through cuttings, revealed both in plan and elevation.

If in plan this intention results in the definition of the irregular polygon that contains the building, in elevation, the cutting becomes evident in the continuous breaks of direction that the shoulder line of the building outlines (Fig. 6).

This operation allows for the reduction of the stadium's scale, lowering its profile in correspondence to the adjoining buildings and granting them the rights of light, at the same time that introduces variation and hierarchy in a facade composition that shows itself to be repetitive and homogeneous.

Simultaneously, and in correspondence to the stadium entrances located in the main squares, as if in a classical gesture, similar to the placement of a pediment, the shoulder line is pulled up. In this way, the entries are pointed out, acquiring identity and importance.

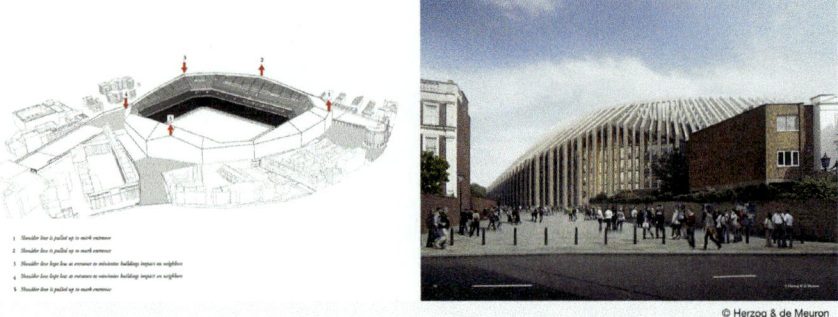

© Herzog & de Meuron

Fig. 6 The shoulder line of the building; West Stand Entrance, Herzog de Meuron (2015), *Stamford Bridge Grounds. Detailed Plan Application. Design and Access Statement* [4, pp. 41 and 34]

Referring to the exterior of the volume, HdM [5] explain that: "(…) the brick piers are a subtle reference to the most important local heritage assets".

Their colour and materiality are a reminder of the architectural qualities of some of the religious buildings located in the conservation areas surrounding the stadium.

It is indisputable that the choice of brick, as a coating material for the Stamford Bridge primary structure, relies on the intent of giving a sense of continuity to London's constructive tradition and to establish a dialogue with the examples of excellence of the city's architecture, especially in what concerns the colour adjustment and the texture exploration of the chosen material.

However, from the point of view of design and geometry, the "skin" that covers the building includes details that reveal a careful consideration of the local circumstances, as well as an interest in constructing the project from the specificities of the site, endowing the building with "roots" so that it can become part of a whole.

We would like to highlight two design themes that witness this particular interest in the local reality.

The brick piers, besides showing up in the facade with alternated different widths, in plan they are radially arranged and aligned by their inner faces, resulting externally in a continuous advancement and retreat along the perimeter of the building (Fig. 7). In a facade where repetition is a necessary theme for the creation of a unitary image, this alternation is fundamental, because it introduces a vibration that enriches the composition giving it density. However, when we look at the characteristic London residential fabric, this theme reappears very often on the facades of the houses marked by the significant saliencies of the bay windows, making us realize that, after all, the game of advancement and retreat of the stadium piers also finds here a reference.

Another interesting aspect in the set of piers that surrounds the building is the fact that they are not just a skin, responsible for the image of the building, but a system with thickness that contains the spectator amenities and that, simultaneously, provides a transition between the exterior and the interior of the stadium.

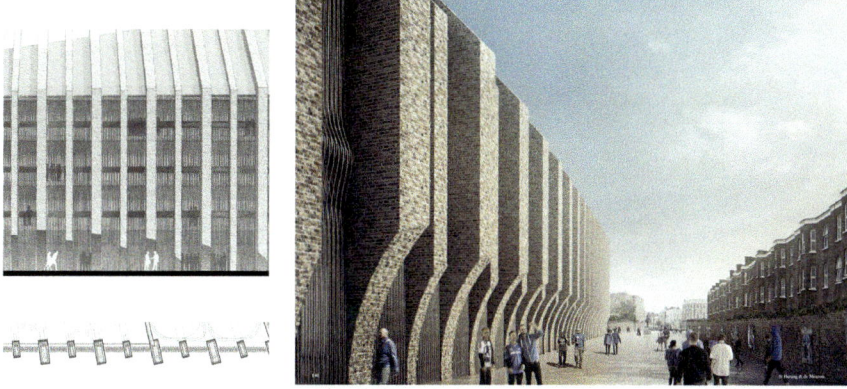

© Herzog & de Meuron

Fig. 7 Detail of the facade—elevation and plan; west elevation and Oswald Stoil Mansions, Herzog de Meuron (2015), *Stamford Bridge Grounds. Detailed Plan Application. Design and Access Statement* [4, pp. 115 and 140]

In this outer ring, in correspondence of three of the five entrances in the stadium (east, south and west) are planned 4–5 storey tall lobbies, carved from the mass of the stadium volume, with the aim of welcoming the spectators (Fig. 8).

In this entrance lobbies, the piers, free at the full height of the building, are topped off with a half-rounded arch that establishes the concordance with the slope of the roof. In the longitudinal sense, the perspective of these "naves" is strongly marked by the continuous succession of the piers and arches, thus acquiring a character

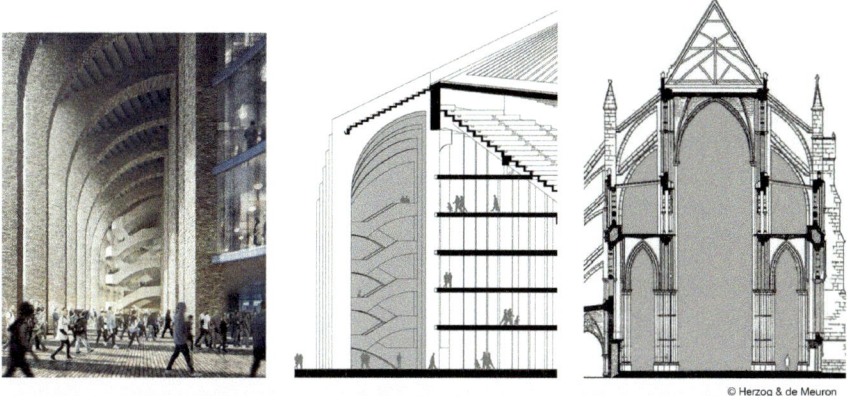

© Herzog & de Meuron

Fig. 8 South-east entrance lobby; west entrance lobby-section; Westminster Abbey-section, Herzog de Meuron (2015), *Stamford Bridge Grounds. Detailed Plan Application. Design and Access Statement*, [4, pp. 131 and 130]

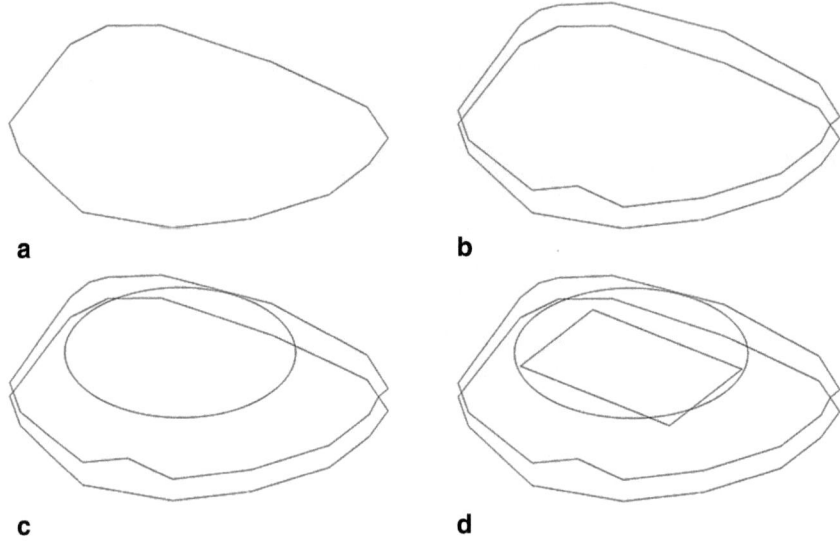

Fig. 9 **a** Irregular polygon; **b** spatial "polyline"; **c** circular ring; **d** rectangle

that recalls, in its scale and proportion, the spatiality of the great English Gothic cathedrals.

From the general point of view of the volume and the surfaces that characterize it, the building can be described as followed (Fig. 9):

- an irregular polygon that defines the base of the building and contains the outermost limit of the stadium volume;
- aligned vertically with this base geometric figure, there is a spatial "polyline" that defines the shoulder line of the elevations;
- pointing out the maximum height of the stadium and aligned vertically with the centre of the pitch, there is a horizontal circular ring;
- in a lower and inner position, that, in horizontal projection, coincides with the limits of the pitch, there is a rectangle concluding the advance of the inner roof of the stadium over the stands.

These four elements define the following surfaces that enclose the exterior volume of the building (Fig. 10):

- vertical planes in the facades (Fig. 10a);
- ruled surfaces, between the spatial "polyline", the circumference and the rectangle:

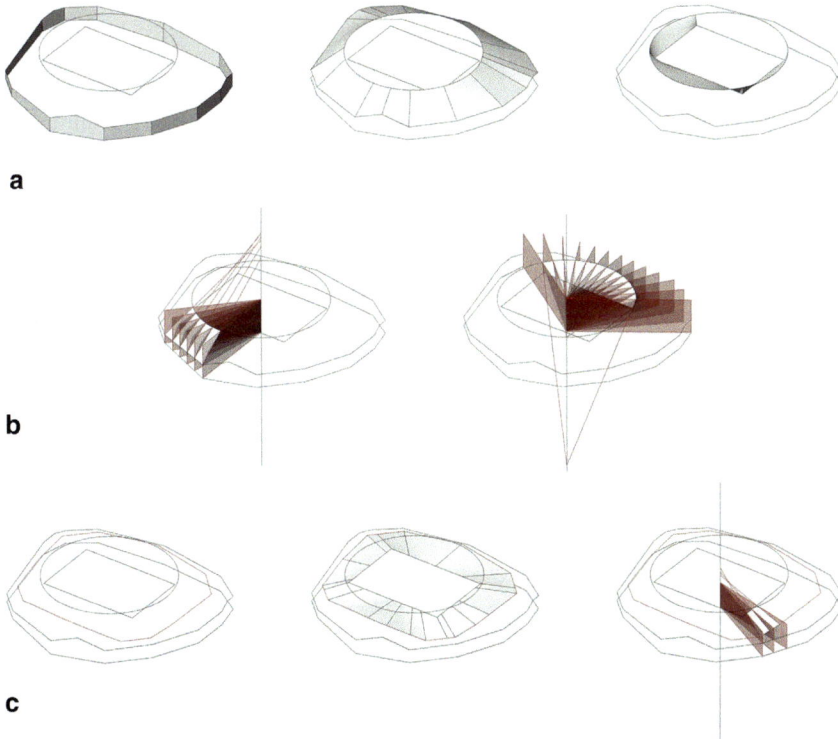

a

b

c

Fig. 10 **a** Vertical plans and ruled surfaces; **b** general conoids; **c** ruled scalene hyperboloids

- each segment of the "polyline" and the arc of circumference, corresponding radially to it, define *general conoids*,[1] since all generatrixes are supported on the straight line directrix, that is normal to the circumference plane and contains its centre (Fig. 10b).
- likewise, each side of the rectangle and the arc of circumference, that corresponds radially to it, also define *general conoids*.

There is also a second spatial "polyline", associated with the upper limit of the stands, that has correspondence with the contour that establishes the separation between the two roof structures of the stadium. Each segment of this "polyline" and the horizontal segments of the rectangle over the pitch define several ruled scalene hyperboloids that give shape to the interior of the roof over the stands (Fig. 10c).

---

[1] As considered by Adrian Gheorghiu and Virgil Dragomir "the conoid is a ruled surface of which the directrices are one curve and two straight lines, either both at a finite distance (general conoid), or one at a finite distance and other at infinity (directing plane conoid)" [6].

2.2 Digital Technologies

Although its description and understanding are relatively immediate, the general shape of the volume is defined by a set of surfaces with a certain geometric complexity that is accentuated from the moment that all of these surfaces are articulated with each other and generate a unitary set, in which any adjustment, however minimal, interferes with the entire volume. This issue becomes even more relevant, since this project is subject to a series of external requests, deriving from the impact of the stadium volume on the surroundings, but also internal, related to the requirements implicit in the design of the stands.

The project was, as such, outlined between the compatibility between the comfort and visibility of the spectators and the reduction of the volume conditioned by the constraints of the site.

As shown in Fig. 11, interior and exterior are closely interconnected, since the geometry of the stands has an impact on the overall geometry of the volume. A change in the steepness of the tiers would modify the volume shoulder line, the overall roof height and the outer mass of the stadium.

The geometry of the tiers was thus generated and controlled with the help of a parametric software, which allowed to manage all these requests, combining the optimal form of the stadium with the best viewing conditions in every seat. For this purpose, it has been defined the value "C" that measures the quality of the sight lines and refers to the ability of a spectator to see the nearest pitch touching the line over the top of the heads of the spectators in front. This value "C" varies between 5 and

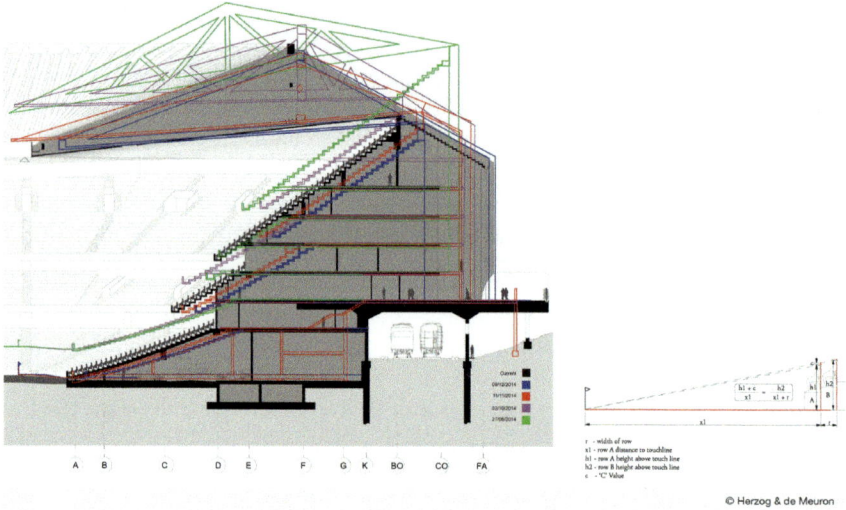

© Herzog & de Meuron

Fig. 11 Stadium section; C-value calculation, Herzog & de Meuron (2015), *Stamford Bridge Grounds. Detailed Plan Application. Design and Access Statement* [4, p. 101]

12 cm and is distributed throughout the stadium, depending on the stand and the category of the seat.

Although the information on this project is still limited, since it is under development, comparing with other HdM works, we can consider that the new digital technologies, such as parametric script, have been fundamental as well as, at least, in the development of these two topics:

• Configuration for the carving of the brick piers;

A cut through cylindrical surfaces, made on the base of the piers with an aesthetic and functional purposes, that happens in specific points of the elevations. Aesthetically, this carving pretends to give a sense of scale to the volume and provide a base to the longest elevations, while, functionally relieving points of greater compression between the building and the site's boundary walls which turns the circulation around the stadium more fluid (Fig. 12a).

• Generation of the compositional pattern applied in the paving of the external public space;

In order to establish a connection with the history of the site and evoke the ancient tributary of the Thames, that ran on the east side of the stadium, HdM took reference from the forms and geology of the waterways to draw the compositional theme of the exterior paving. Through parametric script, the reference pattern can be rationalized, being defined the base geometry that regulates the ulterior compositional alternatives. In this case, the base geometry was intended as specifically organic, varied and dynamic, in order to guide, through its design, the pedestrian flows around the stadium and highlight the building entry points (Fig. 12b).

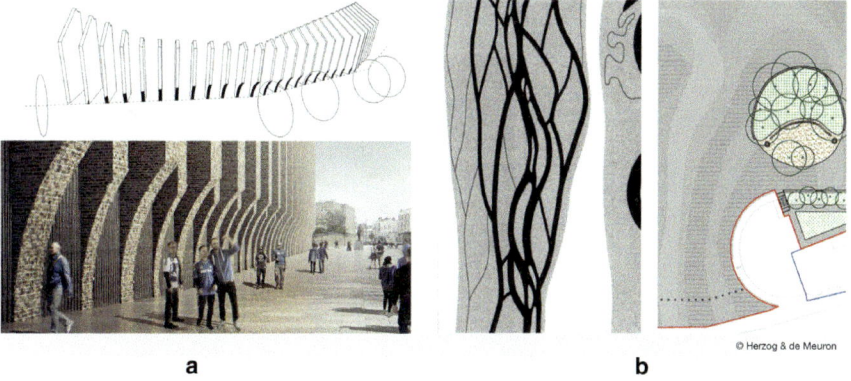

© Herzog & de Meuron

a   b

Fig. 12 **a** Carving on the west facade, Herzog & de Meuron (2015), *Stamford Bridge Grounds. Detailed Plan Application. Design and Access Statement* [3, p. 124]; **b** Flowing river typologies concept image, Herzog de Meuron (2016), *Stamford Bridge Grounds. Detailed Plan Application. Design and Access Statement Addendum* [4, p. 22]

3   Conclusions

In the project for the Stamford Bridge stadium, one of the aspects that stands out most clearly is the fact that geometry is omnipresent in the developed solution, revealing itself in all of its parts, to the point where the image of the building becomes the result of an inextricable interconnection between form, structure and materiality (Fig. 13).

This issue reveals us the importance that HdM give, in the design practice, to geometry, convoking it continually to solve the several project themes and assigning it the role of unifying the building as a whole and articulating the questions of form's design with the functional, structural and constructive aspects.

The new digital technologies are seen as a mean to introduce opportunities for the design, manufacture and construction processes. They are a tool that assists the design, allowing to increase the form's complexity and facilitate the search for alternative solutions, and at the same time they ensure an integrated control of the building in its various parameters and open the possibility to the exploration of new compositional strategies.

However, in HdM's work, the main focus is placed on the design intent and the architectural concept. This means that design tools, analog or digital alike, serve, above all, to translate and clarify the idea and not to create it. When they make use of computing methods, it is not architecture that subjugates itself to the potential of technology, but rather technology that is put at the service of architecture. HdM are not interested in arbitrary geometric possibilities or pure experimentalisms, and oppose to the excessive autonomy and self-reference of architectural objects.

The project for the new Chelsea stadium is, in this regard, paradigmatic.

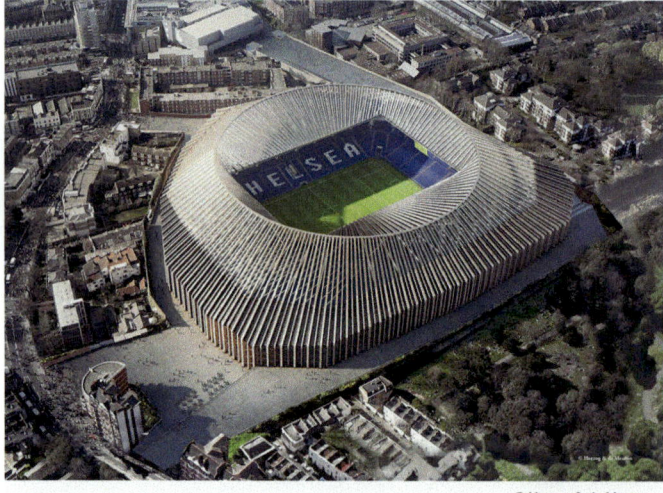

© Herzog & de Meuron

Fig. 13   General view of the new stadium, Herzog & de Meuron (2015), *Stamford Bridge Grounds. Detailed Plan Application. Design and Access Statement* [4, p. 18]

HdM responded to a large-scale theme, that requires, in its typology, the construction of an icon, projecting a building that has a unique and distinctive identity and affirms its character of exception in the city, at the same time that it shows itself to be sensitive to its urban surroundings, seeking to establish with them a conciliatory dialogue.

Once again, it is through drawing that these themes are solved and, above all, with geometry as the key role in decode the existent and bringing into the "visible" world of forms, the "hidden" structures underlying reality.

Acknowledgements The author wishes to thank the office of Herzog & de Meuron for the support.

Assignment co-financed by the European Regional Development Fund (ERDF) through the COMPETE 2020—Operational Programme Competitiveness and Internationalization (POCI) and national funds by the FCT under the POCI-01-0145-FEDER-007744 project.

All the brief extracts of the Detailed Plan Application of the Stamford Bridge Project are reproduced according to the regulations of copyright and the reuse of public sector information mentioned on the London Borough of Hammersmith and Fulham website. "Brief extracts of any material may be reproduced without our permission, under the 'fair dealing' provisions of the Copyright, Designs and Patents Act 1988; for the purposes of research for non-commercial purposes; private study; criticism; review and news reporting—all subject to an acknowledgement of ourselves as the copyright owner."

References

1. "Herzog & de Meuron. From Art to World-Class Architecture", Jacques Herzog interviewed by Hubertus Adam and J. Christoph Burkle (2011). Retrieved Apr 2016 from https://www.herzogdemeuron.com/index/projects/writings/conversations/adam-buerkle-en.html
2. Chevrier, J.-F.: Ornamento, Estructura, Espacio (una conversación con Jacques Herzog). In: Cecilia, F.M., Levene, R. (eds.) 2002–2006 Herzog & de Meuron, El Croquis-129/130. El Croquis Editorial, Madrid (2006)
3. Strehlke, K.: Digital technologies, methods, and tools in support of the architectural development at Herzog & de Meuron. In: ACADIA 09: reForm()—Building a Better Tomorrow. Proceedings of the 29th Annual Conference of the Association for Computer Aided Design in Architecture (ACADIA), Chicago (Illinois) 22–25 Oct 2009, pp. 26–29 (2019)
4. Retrieved in May, 2017 from http://public-access.lbhf.gov.uk/online-applications/applicationDetails.do?activeTab=documents&keyVal=NVL8QGBI0IE00
5. de Meuron, H.: Stamford Bridge Grounds. Detailed Plan Application. Design and Access Statement, p. 35 (2015). Retrieved in May 2017 from http://public-access.lbhf.gov.uk/online-applications/applicationDetails.do?activeTab=documents&keyVal=NVL8QGBI0IE00
6. Gheorghiu, A., Dragomir, V.: Geometry of Structural Forms, p. 166. Applied Science Publishers, London (1978)

Alexandra Castro is an Invited Assistant at the Faculty of Architecture of the University of Porto. She is graduated in Architecture (FAUP, 2002), holds a master in Methodologies of Intervention in the Architectonic Heritage (FAUP, 2009) and is currently developing her Ph.D. research that is focused on the relationship between geometry and contemporary architecture, from the perspective of the impact that digital tools have on architectural practice. Since 2013, she is a researcher at the Centre for Studies in Architecture and Urbanism of FAUP. In 2010, she founded with the architect Nicola Natali the architectural office "Castro Natali".

The Dome as Minimal Housing Unit: "Ghibli" and "D-Home" Prototypes

Micaela Colella

Abstract In this paper are presented two domed minimal housing units: Ghibli and D-Home. Both prototypes were outlined from the discretization of a domed space, using the geometry of the truncated icosahedron, an Archimedean solid that allows the discretization and tessellation of a concentric sphere with only two types of flat surfaces: pentagonal and hexagonal. In this way, both manufacturing and assembly operations will be facilitated and hastened by the presence of only two structural modules. The prototypes are verged in two different constructive solutions—D-Home uses a lightweight timber system, while Ghibli is an experimentation that employs a massive system. The prototypes show the possibilities offered by digital fabrication combined with subtractive and additive manufacturing techniques. The prototypes also show, specifically, how important is for a designer to master geometry and thus be able to create visually complex and unitary solutions, at the same time, reducing costs and facilitating the assembly.

Keywords Dome · Minimal housing · Shelter · Prefabrication · Digital fabrication

1 Introduction

The dome is an architectural shape frequently found as a primitive settlement model in different climatic and cultural areas. We can find it verged in the form of an igloo in the arctic area, where the temperatures are so low that the snow is the most easily available construction material. We can also find it around the Mediterranean area, with different names (e.g. *trulli*, *caprile*, *kazun*) essentially constituted by the same corbelled dome geometry, that is obtained by stratifying the predominant building material, the dry stone. Moreover, it is possible to recognize the geometry of a dome in the earthen homes of African or Middle Eastern villages [1].

M. Colella (✉)
New Fundamentals Research Group, Politecnico di Bari, Bari, Italy
e-mail: mc@newfundamentals.it

© Springer Nature Switzerland AG 2020   29
V. Viana et al. (eds.), *Thinking, Drawing, Modelling*,
Springer Proceedings in Mathematics & Statistics 326,
https://doi.org/10.1007/978-3-030-46804-0_3

It is possible to state that the combination of the most locally available material and the use of the domed shape remain a constructive constant in very different climatic conditions. The primordial need of man to create shelter is thus often explicated in the creation of domed spaces, the only ones that are closed and provide cover in a single constructive act.

For these reasons, domes have never been completely abandoned and were the subject of numerous researches over time. Certainly, among the most significant, it is worth mentioning R. Buckminster Fuller's studies on geodesic domes that were patented in 1945 [2] as structures formed by triangular elements that lie roughly on the surface of a sphere. The geodetic structures allow for the covering of large volumes with minimal surfaces with very high resistance to the weight of the structure. In the 1960s, in the USA, the birth of the hippie movement led many young people to move to areas far from civil society and build simple homes in lightweight structures, with improvised tools. Publications such as the technical manuals *Domebook* [3] and *Domebook 2* [4] explained, in an elementary but exhaustive way, how to build domes based on Fuller's geodesics.

Regarding these themes, two case studies have been developed: D-Home and Ghibli. Both prototypes were conceived from the discretization of a domed space, using the geometry of the truncated icosahedron, an Archimedean solid that allows the tessellation of a concentric sphere with two types of surfaces: pentagonal and hexagonal. In this way, both manufacturing and assembly operations will be facilitated and hastened by the presence of only two structural modules. The prototypes are verged in two different constructive solutions—D-Home uses a lightweight timber system, while Ghibli is an experimentation that employs a massive system (Fig. 1).

2   Ghibli

The regions of the world characterized by extreme climates, such as the African or Arabic desert areas, are conditioned by lack of raw materials to be processed into building materials. While in the rest of the world, a long evolution of architectural forms led to today's building practices, probably these remain as the only places where it is still possible (and desirable) to develop housing strategies deeply linked to the nature of the place and improve significantly the lives of its inhabitants. In this field, the projects built in the African continent by the Italian architect Fabrizio Caròla are very relevant. He uses terracotta bricks to create domes, often ogival, through the use of a wooden compass, made famous by the Egyptian architect Hassan Fathy [5, 6].

The development of new sustainable practices was the reason for this research, that materializes in the construction of an experimental prototype, the "Ghibli Dome", following the Libyan word for warm wind. Ghibli is a domed minimal housing unit, aggregable in villages (Fig. 2).

TRUNCATED ICOSAHEDRON

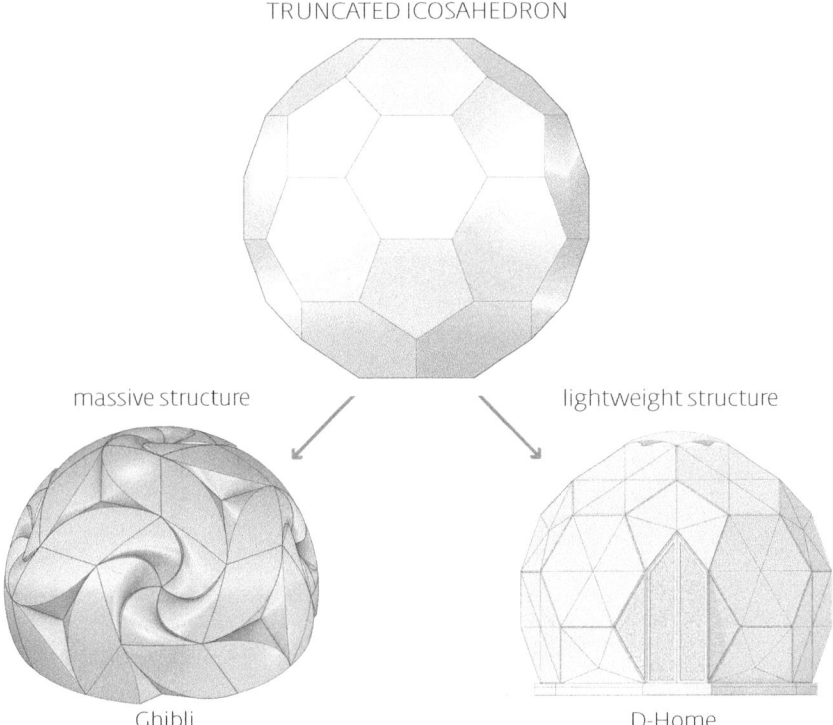

massive structure   lightweight structure

Ghibli   D-Home

Fig. 1 Different structural declinations of the same geometry: the truncated icosahedron

Its construction is possible thanks to a stereotomic assembly of two type-blocks: hexagonal and pentagonal. The intrados of the blocks is optimized using planar faces, thanks to the use of the truncated icosahedral geometry. As such, the geometric construction starts by modelling the basic truncated icosahedron. As previously stated, in order to optimize the construction, it was decided to keep the intradosal faces planar, being these the exact ones of the initial polyhedron, so that a face can be placed on the plane during the construction of the pieces. In order to obtain the extrados of the spherical dome, a spherical portion containing the vertices of the truncated icosahedron has been designed. Then, an offset of the latter spherical surface was accomplished, being the offset size equal to the thickness that was chosen for the dome. Afterwards, the edges of the polyhedron were projected from the centre of the sphere, in order to intersect the extradosal sphere. By doing so, the sphere was divided into equal pentagons and equal hexagons. At this point, a spiral pattern over the dome has been traced, shaping it three-dimensionally, in such a way as to remember the spirals created by sandstorms in the desert (Fig. 3). Its function will be shown next.

A scaled-down prototype of Ghibli has been built (using Lecce's limestone) to test installation issues and thus optimize the realization of the full-scale prototype, during

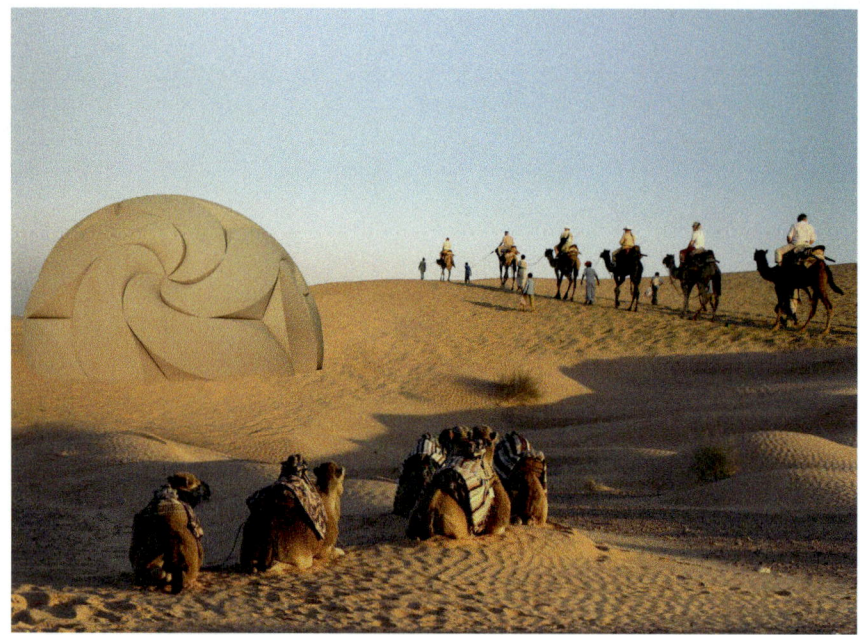

Fig. 2 Rendering of Ghibli

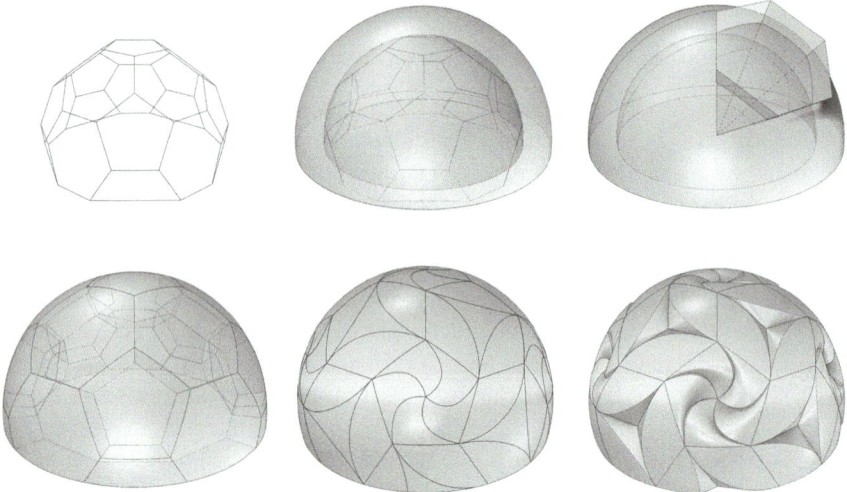

Fig. 3 Geometric construction steps of Ghibli

a didactic workshop in 2015 at the Architecture Department in Rome (Università degli Studi Roma Tre). The stone blocks were obtained by CNC milling. The following issues occurred during the assembly (Fig. 4):

- an unsuitable preparation of the tool paths in the CNC machine by the machining technician caused small but decisive geometric discrepancies compared to the 3D model;
- machining mistakes, compared to the original design geometry, made the blocks impossible to be assembled in a precise manner; consequently, mortar was necessary to fill the wide joints formed between the blocks, an additional procedure that was not planned at project stage;
- despite the reduced size of the blocks, their handling was difficult, because of their heavy weight.

This experience has highlighted several difficulties that may arise in the accomplishment of a construction made of a few heavy blocks. The predicaments of this experience have led to the evolution of the Ghibli project.

Constructively, the new project involves the construction of the dome through the use of a dry process that inserts inert materials inside transparent plastic casings made by rotational moulding of recycled polymers, or also by using 3D printing, once all the necessary stability check-out procedures have been carried out. This way, the constituent elements of the dome, the plastic casings, would be light, and therefore easy to transport and assemble, without the use of heavy transport vehicles and skilled personnel. Among the most suitable materials used for filling, there is certainly sand, that is easily found in many places in the world, even the most inhospitable. The plastic blocks, to be assembled in rows, can be filled with sand only after their positioning, and become stable when the keystone is positioned (Fig. 5).

As it is known, sand has an excellent insulating power due to its high thermal inertia and, as such, it is able to provide good internal comfort conditions, favoured further by the semi-spherical shape of the intrados, that can improve the internal circulation of the air.

Furthermore, recent researches [7, 8] have confirmed that it is possible to induce the sand to a solidification process, through the use of a solution containing urea, calcium chloride and non-pathogenic bacteria (*Bacillus pasteurii*), to turn sand into solid sandstone rock in a short time, to avoid material leakage that could compromise stability, due to plastic breakage or deterioration.

The need for living is accompanied by the need for potable water. The extrados surface of the dome is shaped in carved spirals, that allow conveying of the condensation water on the surface of the plastic blocks, which is created outside, thanks to the high night temperature range. Doing so, water could be collected in an underground storage tank. At the centre of the pentagonal blocks is it expected the presence of a micro-wind system, which would harness the desert wind energy and transform it into electricity to fulfil the most essential human needs. The spherical shape of the

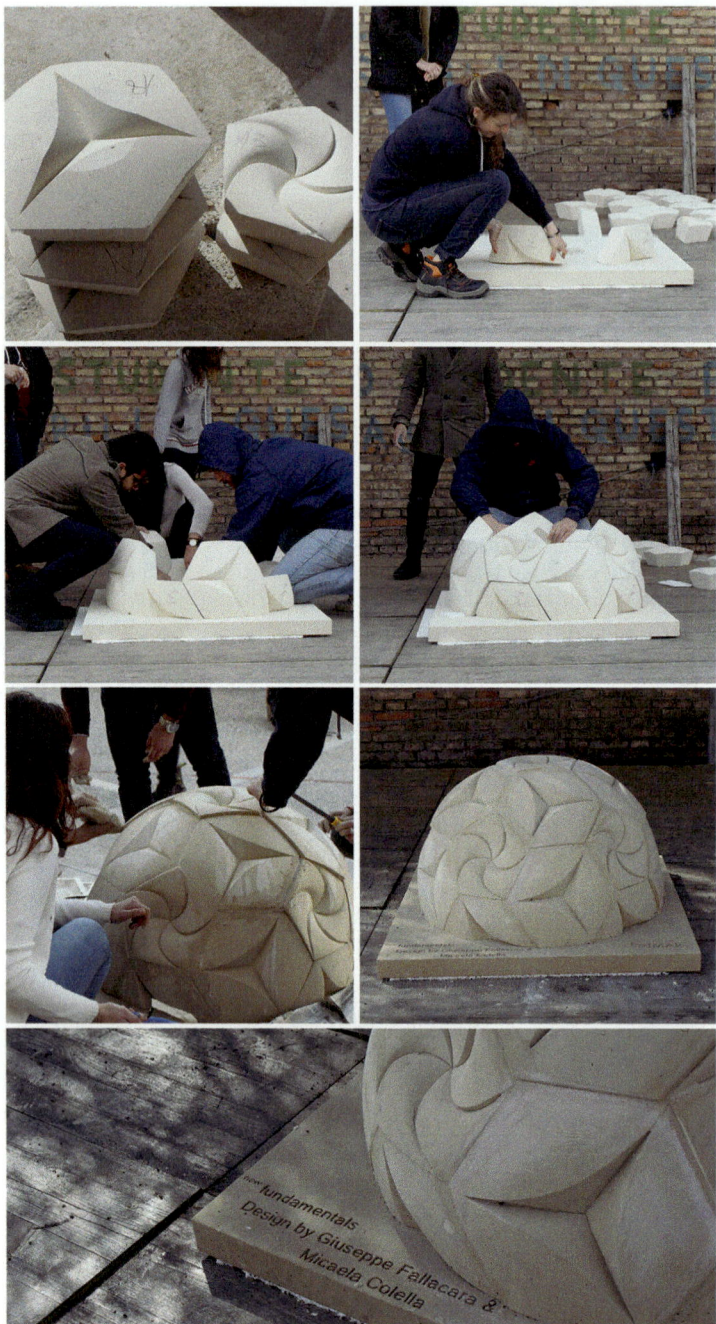

Fig. 4 Assembly of the scaled-down prototype

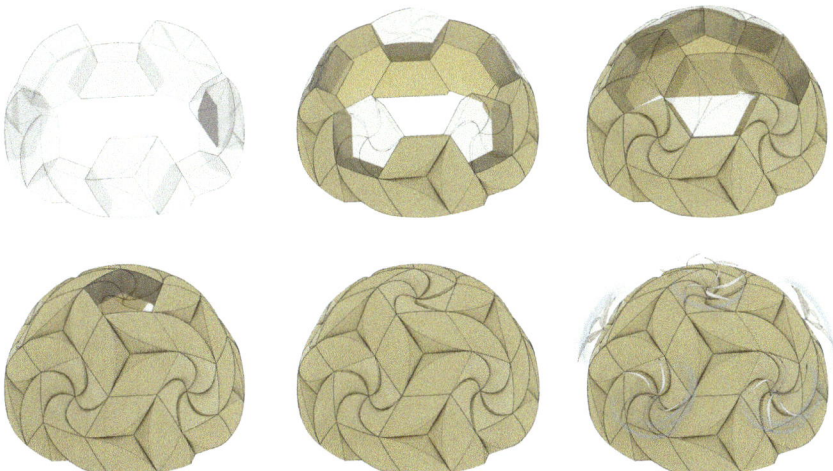

Fig. 5 Three-dimensional simulation of the phases for the dome's assembly

dome and its specific spiral conformation improve the aerodynamics of the construction and allow for a greater uptake of the air currents and, therefore, better energy plant performance.

3 D-Home

The second prototype, "D-Home", was conceived to accomplish a small architectural project deeply linked to the architectural and cultural heritage of the Mediterranean basin and to create minimum temporary shelters for camping and touristic purposes (Fig. 6).

D-Home is a small wooden prefabricated dome with a total size of ten square metres. Similarly to the previous example, it was outlined from the discretization of a dome by a truncated icosahedron, consisting of hexagonal and pentagonal faces broken down into triangles (Fig. 7). Each triangular module is constituted by a closing panel, three lamella beams (20 cm height) placed on the edge of each module, and an additional lamella with three beams (*Y*-shaped) placed at the centre of the triangle, with bracing function (Fig. 8). The inner beams have semi-circular cuts that are useful to reduce the weight of the structure, without compromising its static stability. The aesthetics of the form ideally resembles the ordered complexity of Arabic architecture (e.g. *muqarnaṣ*). The decomposition in triangular modules facilitates the accomplishment of assembly operations, thanks to the use of modular wooden elements. The wood has been preferred to other materials for its lightness, affordability and high degree of prefabrication.

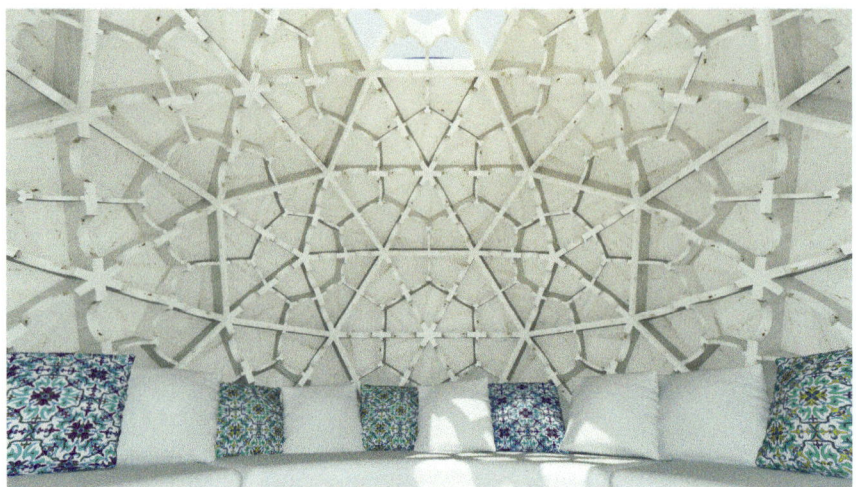

Fig. 6 Interior view of the D-Home

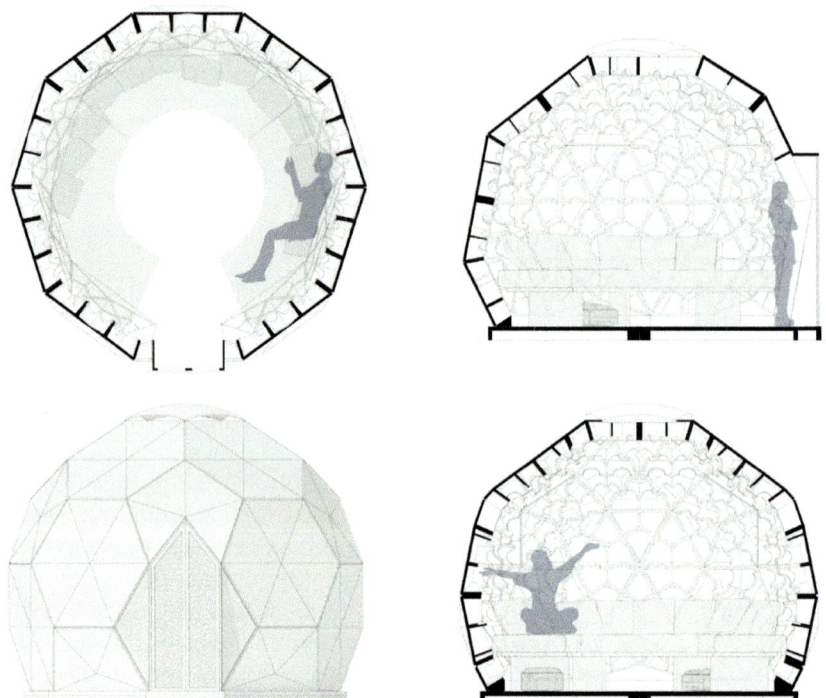

Fig. 7 Plan, sections and elevation of the D-Home

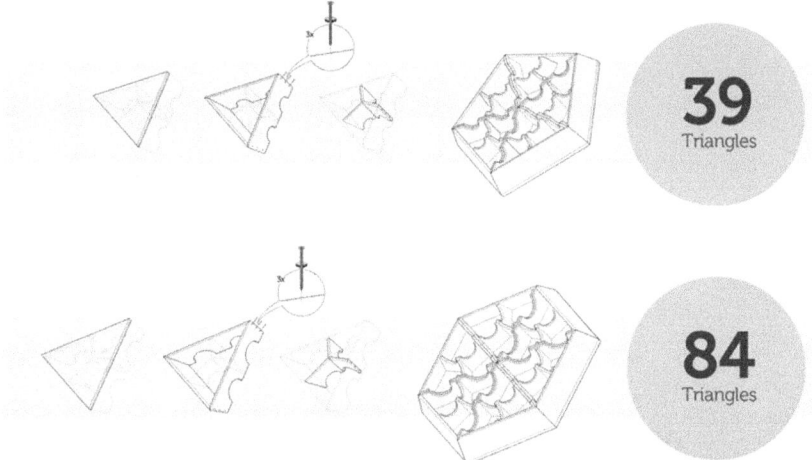

Fig. 8 Wooden components of the D-Home

Despite the complexity of its aesthetic outcome, the D-Home is installed in a very simple way, simply by fixing panels to each other, by means of self-tapping steel screws. All wooden components may be previously shaped using numerical control machines or fablab tools.

After the assemblage, all the joints are sealed with acrylic silicone to waterproof the shelter, and a subsequent white enamelling completes the finishing. White colour is naturally chosen for its ability to reflect the sun's light, thus avoiding overheating of indoor environments. For this reason, white has been the dominant colour of buildings in the Mediterranean area over time (Fig. 9).

Despite their small dimensions, the temporary use and the necessary affordability of these structures, certain design choices have been adopted in order to ensure the indoor comfort in summer heat conditions (the season for which its prevalent use was imagined). Natural ventilation is improved by an opening placed at the top that, although protected from rain, allows the heat to go outside. Even the entrance improves natural ventilation, thanks to a perforated door, inspired by the principle of *moucharabieh*, an Arabic natural ventilation device, whose intricate geometry can be easily made with CNC milling.

A real-size prototype of a hexagonal module has been realized (Fig. 10), in order to verify the feasibility of its manufacturing. This experience has confirmed that the only element of relative complexity is found in the uneven angle between the various modules required to create the global geometry.

Future developments could relate the application of the D-Home constructive system to larger housing solutions, requiring higher comfort requirements. In fact, it is possible to easily improve the insulation performance of a housing module with

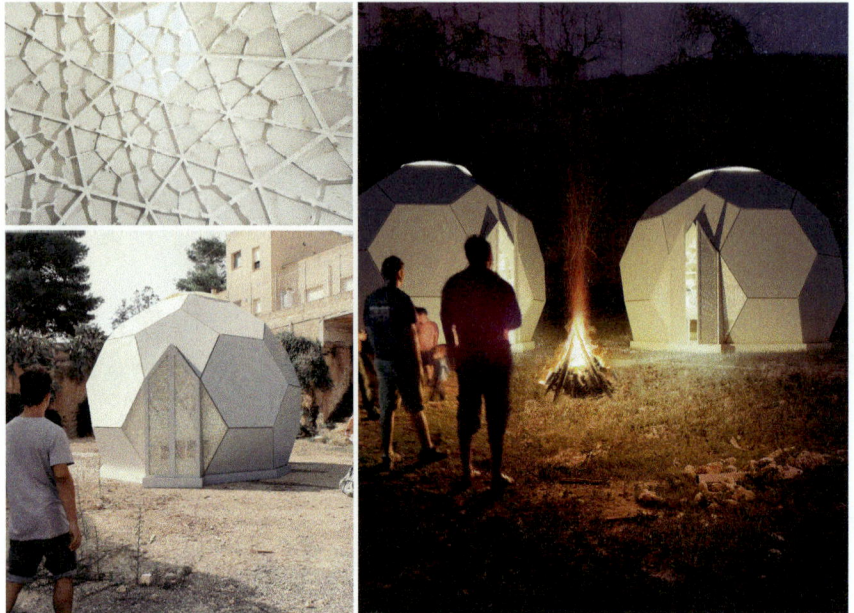

Fig. 9 Exterior views of the D-Home

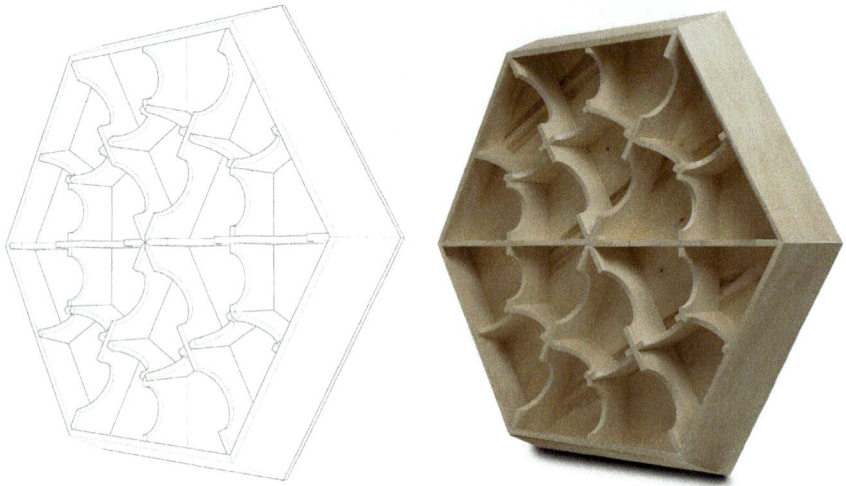

Fig. 10 Real-size prototype of a hexagonal module

appropriate insulation panels. In addition, further reflections should be made on the positioning of windows that could ensure more adequate lighting, ventilation and security conditions.

4 Conclusions

Both prototypes described in this contribution specifically show how important it is for a designer to master geometry. Geometry allows us to create visually complex and unitary solutions at the same time, and its knowledge enable us to build our projects discretizing whole surfaces into more simple modules, thus reducing costs and facilitating the assemblage. This is particularly true for structures such us geodesic domes resulting from geometric projections and tessellations. Another important aspect highlighted in this paper is the fact that, in spite of very different climatic conditions, the construction of domed houses built using the materials available on site tends to remain an invariant. In response to the primordial human need of creating shelter, domed spaces are the only ones able to enclose and cover in a single circular constructive movement.

Acknowledgements The projects are part of Colella's doctoral dissertation, and they were developed under the scientific supervision of Prof. Giuseppe Fallacara. Colella and Fallacara are also co-designers of Ghibli and D-Home prototypes. Finally, the author would like to thank the companies Pimar and WoodTec for the fabrication and construction of the prototypes.

References

1. Mecca, S., Dipasquale, M. (eds.): Earthen Domes and Habitats. Edizioni ETS, Pisa, Italy (2009)
2. Fuller, R.B.: Building construction (patent n. US2682235 A) (1954). Retrieved in May 2016 from: https://www.google.com/patents/US2682235
3. Kahn, L.: Domebook. Pacific Domes, Bolinas (1970)
4. Kahn, L.: Domebook 2. Pacific Domes, Bolinas (1971)
5. Caròla, F.: Vivendo, pensando, facendo… memorie di un architetto. Napoli, Intra Moenia, Italy (2005)
6. Alini, L. (ed.): Cupole per abitare. Un omaggio a Fabrizio Carola. Libellula Edizioni, Tricase (2012)
7. Larsson, M.: Dune: arenaceous anti-desertification architecture. In: Macro-engineering Seawater in Unique Environments, pp. 431–463. Springer, Berlin Heidelberg (2010)
8. Dosier, G.K.: Methods for making construction material using enzyme producing bacteria (patent n. US 8728365 B2) (2011). Retrieved in Feb 2016 from: https://www.google.com/patents/US8728365

Micaela Colella is an architect and Ph.D. in architectural design. Her research focus is on the advances in prefabrication systems that use natural materials for sustainable houses in the Mediterranean area. She is a partner and co-founder of "New Fundamentals Research Group" directed by Prof. Giuseppe Fallacara as well as co-founder of the architecture firm "Barberio Colella ARC".

Geometry and Art

Lino Cabezas Gelabert

Abstract The relationships between art and geometry were not always the same, and this might be explained from the great differences between artistic styles and the way in which art has been socially considered, that profoundly changed through the centuries, alongside the development and evolution of the knowledge on geometry. The complexity of these relations may be found in the history of geometry itself, as well as in the structure and the way in which the professional and academic institutions work, not only in regard to their practice, but also to their learning systems. On this subject, a number of examples are considered to demonstrate the connection between the knowledge on geometry and its incidence or application in the works of art. One of these refers to the representation of the human figure, since the proposals to establish the standards of beauty span from the Classic Greece, until the development of contemporary anthropometry.

Keywords Art · Geometry · Descriptive geometry · History of representation

1 A Historical Debate

The art–geometry binomial has been the subject of many interpretations over the years. First of all, due to the conceptual evolution experimented by both terms, there are large-scale episodes regarding this relation, although sometimes under other denominations, such as the general art–science relationship. Much the same way, the art–science binomial played a crucial role in one of the most important academic debates prompted by the confrontation between the prescriptive conceptions of art and the subjective and intuitive arts approaches. This discussion on the functions of geometry, in this regard, has been constant, particularly in such important questions as the human body's representation and proportions (Fig. 1).

L. C. Gelabert (✉)
University of Barcelona, Barcelona, Spain
e-mail: linocabezas@gmail.com

© Springer Nature Switzerland AG 2020   41
V. Viana et al. (eds.), *Thinking, Drawing, Modelling*,
Springer Proceedings in Mathematics & Statistics 326,
https://doi.org/10.1007/978-3-030-46804-0_4

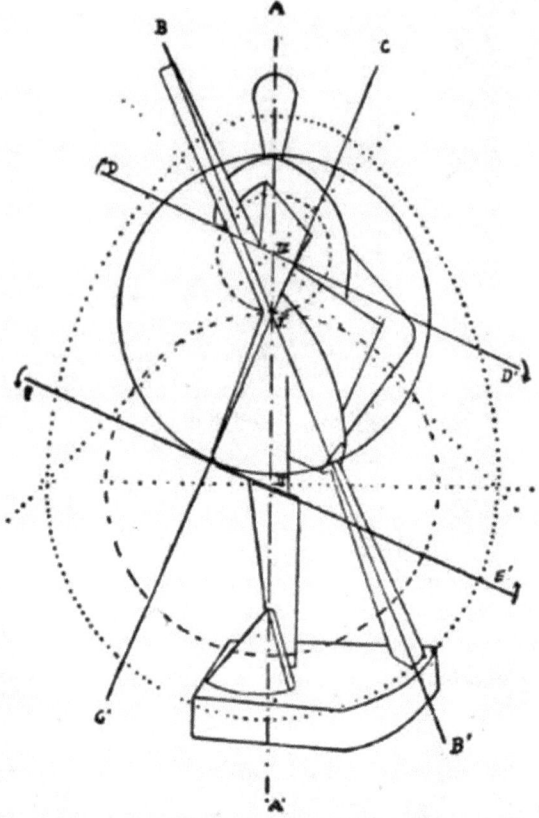

Fig. 1 Vantongerloo, *Le Gondolier*, 1817

Geometry has caused controversies within the art world, particularly, within academic institutions, where art´s functions and significance are frequently debated. This idea can be illustrated by the famous report on arts in general, and paint´s study in particular, written by Goya in 1792 and sent to Bernardo de Iriarte, vice protector of the Real Academia de Bellas Artes of San Fernando (Madrid), that trivialized the teaching of geometry in the Academy, stating that

> Nor should a time be predetermined that students study Geometry or Perspective to overcome difficulties in drawing, for this itself will necessarily be demanded in time for those who discover an aptitude and talent, and the more they advance in these subjects, the more easily they attain knowledge in the other Arts.[1]

With these words, Goya contradicts the prevailing view among scholars of his time, exemplified by the report of Mariano Salvador de Maella defending the current

---

[1]Goya's letter has been published within the following exhibition's catalogue: *Renovación. Crisis. Continuismo. La Real Academia de Bellas Artes de San Fernando en 1792*. Madrid: R. A. de BB. AA. de San Fernando, 1992.

academic system, which ordered that each student could only attend his drawing course after having completed the geometry course [1].

At that time, geometry, perspective, and anatomy were pleaded as the scientific backbones and obligatory underpinnings of any Academy of Fine Arts, thanks to the scientific endorsement given by some important institutions, such as the Academy of San Lucas of Rome, the Clementine Academy of Bologna or the Academy of Painting and Sculpture of Paris. In line with these institutions, the Royal Academy of Fine Arts of San Fernando in Madrid established in 1766; at the behest of the illustrious sculptor Felipe de Castro, these three chairs with the explicit support of Antonio Raphael Mengs, first court painter of Spain.[2]

Nevertheless, as Goya himself knew, there was a distinct shortage of books specialized in teaching these disciplines, which were, most of the times, scarcely known by the barely educated teachers.[3] Then, as much as today, the need of quality books, as well as trained teachers, arised as a key pillar for the art´s teaching qualitative improvement.

Nowadays, many art education professors have doubts about the functions and motivations of introducing geometry into the syllabus or, most particularly, the role that geometry plays in drawing as an artistic discipline. Furthermore, many people have discussed geometry as something accessory to the world of arts, whose specific interests largely differ from those of geometry, which they find more appropriate for scientific disciplines.

Regarding this issue, we should avoid using, mainly, historical arguments to assert that geometry has always been connected with arts. Indeed, throughout art history, there are plenty of examples in the form of testimonies defending the artistic relevance of geometry, such as the contributions of Vitruvius in Classical times, Dürer in Renaissance or Le Corbusier in Modernity (Fig. 2).

To help answer all these questions, the treaties that address the relationship between geometry and art can be clarifying instruments and, as such, one fact should not be overlooked: as part of the work of any current teacher, written sources must be studied and known. On the other hand, teachers and professors must be kept up to date, so that they are able to convey to students the same curiosity and critical attitude contained in these texts.

Against any misunderstandings, it can also be said that the assertion that geometric knowledge is stabilized and that it has been concluded in the past, cannot be stated as entirely truthful. According to historical testimonies, the contents of geometry, much the same way as art, have not remained unchanged, which contradicts the

---

[2]The best study on the foundation and the first years of the Real Academia de Bellas Artes of San Fernando is still the French Hispanist Claude Bédat's doctoral thesis, published in French in 1973 and translated into Spanish for the edition quoted in the previous footnote.

[3]In a well-documented work by Alicia Quintana Martínez, about the San Fernando Academy, a part of it is dedicated to the publication that collects the "imparted subjects. The development of textbooks". There, testimonies about mediocrity and the frustrated commitments pursued by the editorial policy of the Academy are described and analyzed in Quintana Martínez, Alicia. *La arquitectura y los arquitectos en la Real Academia de Bellas Artes de San Fernando 1744–1774.* Madrid: Xarait, 1983, pp. 64–75.

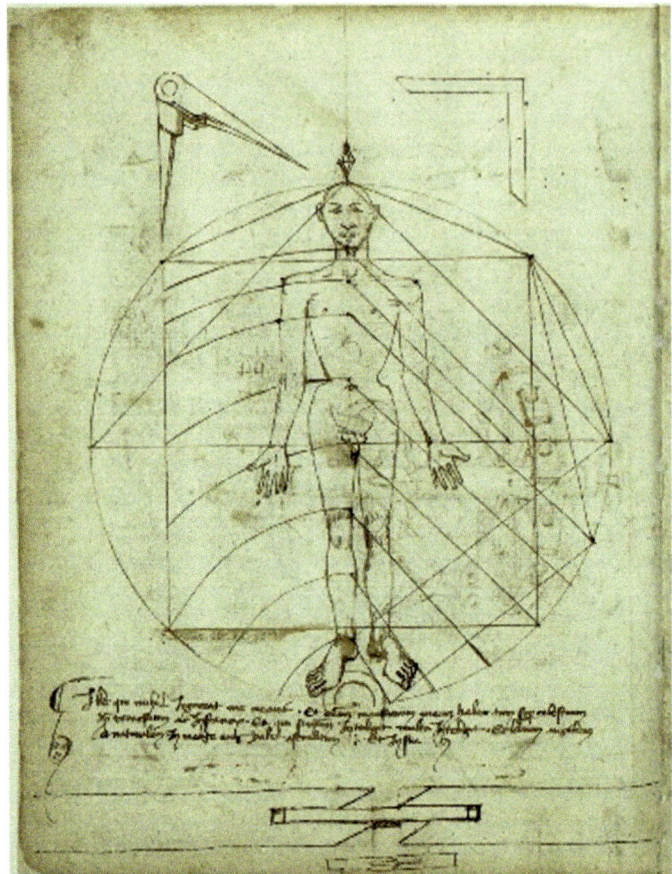

Fig. 2 Taccola, *De ingeneis,* 1419–1433

idea that these contents do not have the capacity to adapt themselves to scientific, technological, or cultural evolution.

2 Geometry and Geometries

Nowadays, it is difficult to talk about geometry without using adjectives. Traditionally defined as a scientific discipline that rigorously studies the concept of space and the figures that can be imagined occupying this space, geometry can be termed in different ways: Euclidean, analytical, metric, descriptive, projective, plane, three-dimensional solid, kinematics, differential, geometry of *n* dimensions, and many others. Between these terms, some synonyms are used, as in the case of classical geometry, also known as Euclidean or intuitive geometry.

On the other hand, geometry has also been used to term/characterize some art types: the so-called geometric styles, which include geometric abstraction as well as geometric ornamentation, both studied in the history of applied arts. In a similar way, it is also used as an adjective to qualify a particular type of drawing: geometric drawing.

Despite the remote origins of the foundation of geometry, most of its branches come from recent historical formulations, especially those developed throughout the eighteenth and nineteenth centuries. In the former, the analytical, differential, projective, and descriptive geometries are consolidated or formulated and, in the latter, the so-called non-Euclidean geometries.

Although a historical in-depth analysis is not possible here, it should be noted that the non-Euclidean geometries of the nineteenth century, despite their scientific importance, have only an indirect, or almost null, connection with the arts. It is not in vain that the non-Euclidean or modern geometries, fundamentally the elliptical or Riemann's and the hyperbolic geometry of Lobachévski, are also known as non-intuitive, because of the impossibility of their representation with sensitive space's figures.[4]

Here, we can address the broad field of geometry regarding its relations with the art world, as well as in relation to the theoretical training of artists through texts related to arts. This limit tightens the vast panorama of geometry to a smaller field, which one could, provisionally, call graphical geometry. However, to address this issue, it is necessary to get rid of some usual preconceptions.

The academic chasm between artistic and scientific culture imposed a rigid verge between both, this being the cause of some prejudices based on weak conceptual or historical arguments. Faced with this dilemma, geometry has been required to position itself as a scientific or philosophical question, somehow disconnected from the art world. On the other hand, if geometry pretends to be linked to art, it aims to assume an exclusively instrumental role. For all these reasons and according to this belief, some consider geometry as a subject that can be dispensed from the interests of art without losing anything essential to it.

3   The Art of Geometry

During the Middle Ages, it can be noted how practical geometry related to craftsmen and how, known by the Latin term *geometria fabrorum*, it was clearly separated from the theoretical corpus of which it was a part of, along with the arithmetic, astronomy, and music, of the *quadrivium* of the liberal arts. Likewise, the *quadrivium*, together with the *trivium* (rhetoric, grammar, and dialectic) constituted the system of scholastic

---

[4]Non-Euclidean geometries are primarily associated with names like Carl Friedrich Gauss (1777–1857), Nicolai Ivanovich Lobachévski (1792–1856) and Janos Bolyai (1802–1860), without forgetting pioneers like Girolamo Saccheri (1667–1733), Johann Heinrich Lambert (1728–1777) or Adrien Marie Legendre (1752–1833).

Fig. 3 Villard de
Honnecourt, *Cuadeno*, s.
XIII

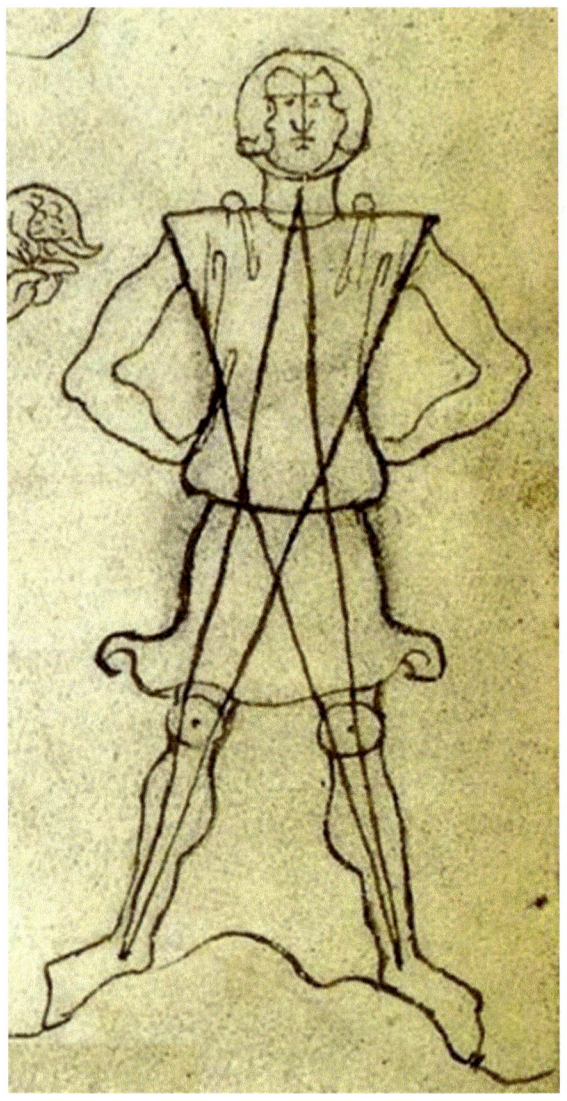

teaching in medieval universities.[5] As an example of the use of practical geometry
during the thirteenth century, in the well-known notebooks of master builder Villard
de Honnecourt, these words can be read: "Here begins the method of drawing portraits
as taught by the art of geometry, to facilitate working. And in the other sheet, we can
find the methods of the masonry" [2] (Fig. 3).

---

[5] An interesting and well-documented article in where this question is approached is Ruiz de La
Rosa, José A." Geometriai fabrorum" o la antítesis de las teorías sofisticadas" in Boletín Académico
nº 7, octubre 1987.

In fact, during the Middle Ages, multiple interests and independent geometric knowledge coexisted: that of the scholars and that of the artisans, in a relationship that also exists between the two of them. However, their origin was not very different: the human being's ancestral ability to recognize and compare different shapes and sizes that, over the centuries, will build a body of knowledge aimed at classifying and studying the extension, position, and shape of objects.

After these considerations of art and geometry, theoretical elaborations can help to better understand the framework of the general learning of the arts and its relationship with the institutions dedicated to them. In this respect, we must bear in mind the specific sources. Nonetheless, in a narrow sense, we can only speak of geometry books after the invention of the printing press in the middle of the fifteenth century. However, the foundation of the geometric knowledge collected by treaties is previous, going back to the origins of History.

4   The Theoretical Foundation of Geometry

Without being able to be defined as a textbook, although sometimes used as such, the great work of Euclid, *The Elements of Geometry* (300 BC) is considered by many, along with the Bible, as the most read book in History, a best-seller for more than twenty centuries.[6]

Its origins in Greek thought are born with the awareness that geometric knowledge cannot be obtained only through sensitive experience within the physical world, but can be deduced by linking thoughts in a logical sequence. The axiomatic method of geometry was then formulated and the *Elements* would be established as their best and greatest exponent. This work was converted, first in the Arab world and then by scholasticism, into a book adaptable to the uses of each culture and able to assume a hegemonic position in the future, which has been maintained until the middle and university education of the nineteenth century, where it would, consequently, end up losing much of its secular privilege.

The culmination of the axiomatic method of the *Elements* is usually dated with the publication in 1899 of the *Grunlagen der Geometrie* (Foundations of Geometry) by David Hilbert, where this German mathematician established five groups of axioms thinking in logical and non-intuitive terms [4].

Critical revisions of classical geometry by mathematicians were not exempt from passion either. In the year 1959, the cry "Down with Euclid!" was heard in a mathematics congress held in Royaumont, north of Paris. This anathema, pronounced by the Frenchman Jean Dieudonné, was not directed against Euclid himself, but against a classical and conservative conception of the teaching of mathematics, which, in his opinion, was not adapted to the evolution of knowledge. However, the key to the

---

[6]The *Elements*, on the contrary of what happens with other languages, has not received a complete translation into Spanish until very recent times. This circumstance was compensated with the quality and care of the three volumes edition published in [3].

historical success of Euclid was, beyond identifying his name with a mathematical discipline, in naming a method of axiomatization. In this sense, this precision is very clear: "Euclid, for the many generations that have been nourished by its substance, may have been less a professor of geometry than a professor of logic" [5] (Fig. 4).

A historical anecdote gives an accurate idea of the capacity of the *Elements* to become a powerful instrument of argumentation and logical reasoning. After the establishment of the Jesuit mission in China in the late sixteenth century, the need to achieve maximum effectiveness in evangelization through images led the Jesuits to launch a program of scientific acculturation. Matteo Ricci, founder of the mission in 1583, in his strategy, in the year 1606, had the first six books of the *Elements* of Euclid translated into Chinese by his disciple Paolo Sui. The translation, according to a testimony of the time, was for the Chinese "a thing of great wonder, in this way never seen before, the type of book and the way to prove and demonstrate with such evidence" [6]. This would allow the translation in 1729, also into Chinese, of the *Perspectiva pictorum et architectorum* by Andrea Pozzo.

Fig. 4 Ramón Llull, *Liber de geometria nova et compendiosa*, 1036

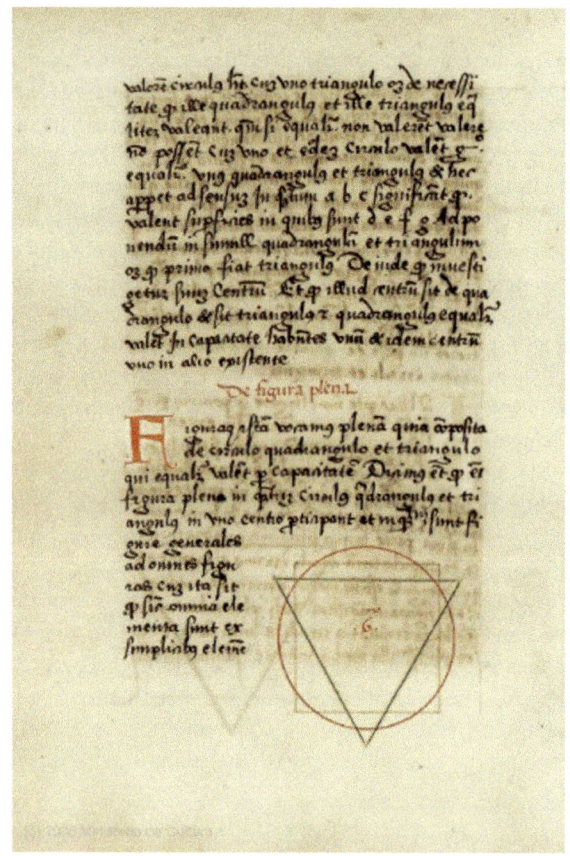

5 Scienctia Scientiarum

The years of the first Chinese translation of the *Elements* of Euclid were the first of that century, later known as the century of the Scientific Revolution. During this period, geometry assumed the role of "*scientia universalis*" or "*scientia scientiarum*" (science of all sciences), being considered as the vehicle for access, par excellence, to all levels of human thought and activities, with arts included, obviously.

It must be recalled that Desargues, founder of projective geometry, would see his work published thanks to his friend and follower Abraham Bosse, first Paris Academy Perspective professor, who was expelled for defending the "truth" of his scientific knowledge about this discipline (Fig. 5).

In this context, some facts are explained, such as the structure of the work of the philosopher Spinoza (1632–1677), entitled *Ethica, ordine geometrico demonstrata,*

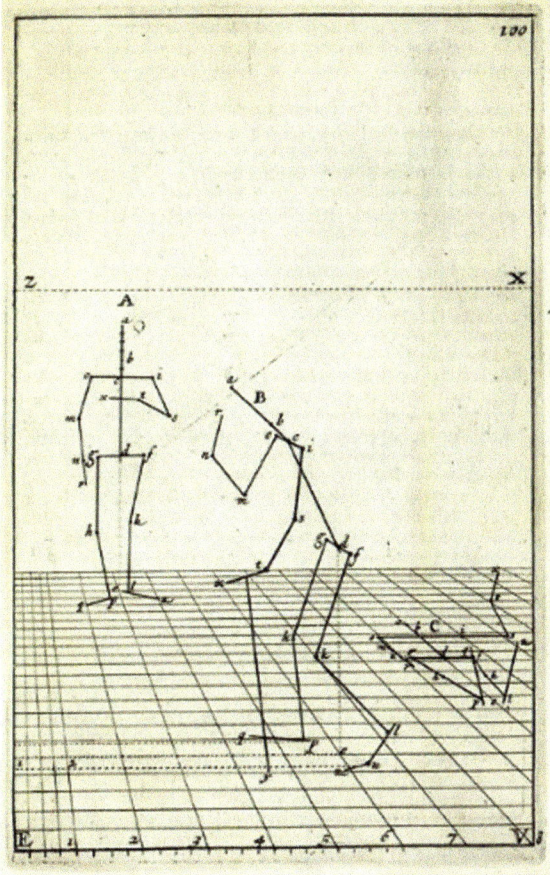

Fig. 5 Abraham Bosse, *Maniere Universelle de Mr Desargues...*, 1648

which, like his *Principles of Cartesian philosophy* are more *geometrico demonstratae*, that is: proven geometric mode from definitions, following with axioms, to arrive at the propositions from which the corollaries can be followed. In short, the philosopher proceeds with the same deductive method of the *Elements*.

The importance given to the logical-deductive reasoning of geometry was evident since Ancient Greece times, expressed in the maxim of the Platonic Academy (fourth century BC), "Let no-one ignorant of geometry enter here", understood in the context of Pythagorean and Platonic thought looking for an ideal philosophical order expressible in arithmetic systems and in the comparison of geometric figures.

6 Texts About Geometry

The popularization of geometry knowledge was made possible, among other circumstances, thanks to the invention of the printing press. In the same way, the abandonment of Latin in favor of vernacular languages meant a profound change in the new humanistic institutions and technical interests, a fact that would provoke some reactions. Rodrigo Zamorano, as the first translator of Euclid's work into Spanish, is forced to defend his ideas along with the very same pages of his translation of the *Elements* against the opinions of his time:

> Seeming better to me, the advantage that some took from the murmuring that I forcedly have to suffer from others, it seems to them, that the stepping of science into everyday language is to make them something mechanical, not seeing that the authors who wrote them at the beginning, left them written in a language that was then as vulgar as now ours is [3, sheet 7v].

This justification would not be necessary a few years later, since the translation of Euclid's *Optics* gained the license and privilege of King Felipe II, who authorized, in 1584, its publication under the title *La Perspectiva y Especularía de Euclides traducidas en vulgar Castellano por Pedro Ambrosio Ondériz*, where the very same translator (Ondériz) reminded us "the reason to do it was that, as Your Majesty ordered in this, your court, Mathematics shall be read in Castilian language" [7].

7 Practical Geometry

With the appearance of the printing press, parallel to the edition of the classical texts, the geometrical knowledge inside medieval workshops which, until then, had been disseminated thanks to oral transmission within the different guilds, will be published. The geometry of the medieval craftsmen had been gestated as something separated from any theoretical reflection and their purpose was, besides solving immediate construction problems, to generate a wide repertoire of forms based on the construction of geometric diagrams applicable to different trades. The survival of those methods was possible thanks to the fact that some artists decided to publish the secrets of their professions at the dawn of Renaissance. The rudimentary aspects of the practical geometries related to different trades justify that, even nowadays, it remains pejoratively referred to as "tailor's geometry" referring to its practical and elementary character. This name would become the most adequate, and historically endorsed. Indeed, the *Libro de Geometría, Práctica y Traça. El qual trata de lo tocante al officio de sastre* (1580) by Juan de Alcega is not just an occasional publication. This work, which had the approval of that time's tailoring masters, was reissued and served as model for later ones.[7] *Geometría y Traça para el oficio de sastre* edited in Seville in 1588 [8] followed this publication, which was equally followed by other texts dedicated to the same subject.[8]

All these practical geometries would, subsequently, reach the art treaties which could somehow be named by us as "cults", collected to a greater or lesser extent, as can be seen in those of Albrecht Dürer or in the treatise of the Bolognese architect Sebastiano Serlio, that will be mentioned in the next lines (Fig. 6).

8 "I Will not Speak as a Mathematician but as a Painter"

It is usually stated that the artistic literature of the Modern Age begins in the Italian Renaissance with the writing, in the year 1435, of the treaty *De pictura* by humanist Leon Battista Alberti. This work represents the beginning of a new genre and marks the guidelines of a large part of the subsequent treaties. In relation to geometry, Alberti's text is the first, in absolute terms, that deals with perspective while exposing an operational procedure for painters, based on the geometric reduction of human vision, following the theoretical models of optical science.

As has been noted in more than one occasion,

> In fact, the perspective procedure introduced a new factor—the incorporation of the 'geometric science' in the work of painters, something completely alien until then to the artisan workshops, that was charged with enormous consequences, because it was part of the criteria that states that the artistic activity had to be conducted according to a genuine theory [9].

This circumstance explains why Alberti begins by defining elementary geometric concepts, starting with points, lines, surfaces, and other geometric figures to, after explaining the basic concepts of optics, reach out to propose his famous definition of painting as "flat section of the visual pyramid" (Fig. 7).

Therefore, we are able to understand the meaning of the first words of the writing in the text:

---

[7]Context and references to the publication of this work have been studied in López Piñero, José María (1979) *Ciencia y técnica en la sociedad española de los siglos XVI y XVII*. Barcelona: Labor, p. 176.

[8]Among other titles, we can quote: La Rocha: *Geometría y traza... de sastre...*, 1618; Anduxar: *Geometría y trazas pertenecientes al oficio de sastre*, 1640.

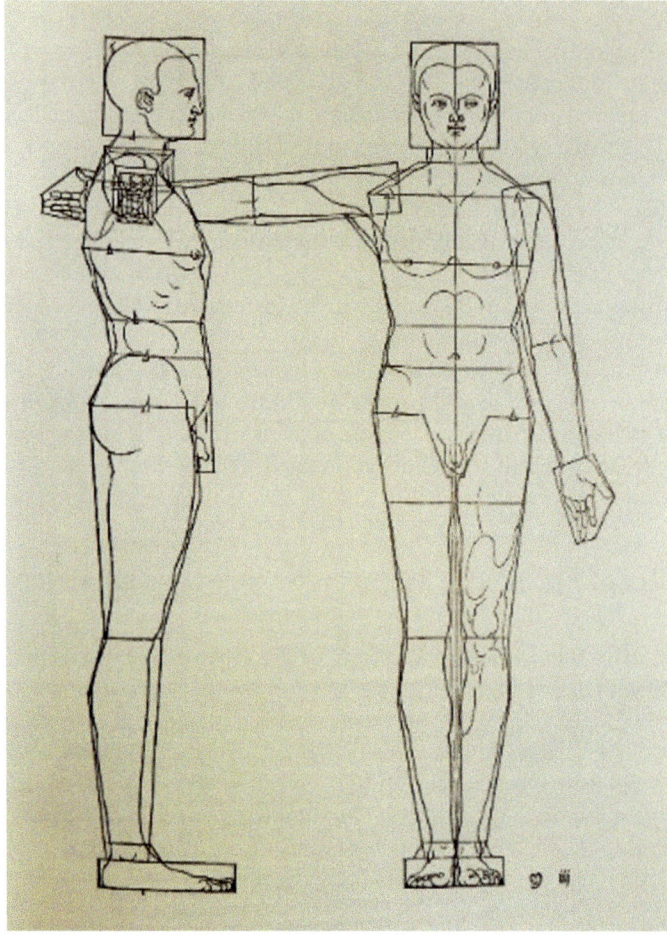

Fig. 6 Albrecht Dürer, *Von Menschlicher Proportion*, 1928

> Having to write about Painting via these brief comments, I will take from Mathematicians,
> to make myself understood more clearly, everything that leads to my subject. This being
> understood, I will explain the best thing that Painting can be, following the same principles
> of nature. But in my speech, I gave a warning, that I would speak not as a Mathematician
> but as a Painter [10].

From this moment on, perspective texts mark out a history of five hundred years in
something that can be considered as the most important contribution to geometry
made by plastic artists.

The first printed text dedicated exclusively to pictorial perspective, *De artificiali
perspectiva*, is written by the Frenchman Jean Pélerin, called Viator, a text reissued
on multiple occasions thanks to its practical approach designed for the needs of

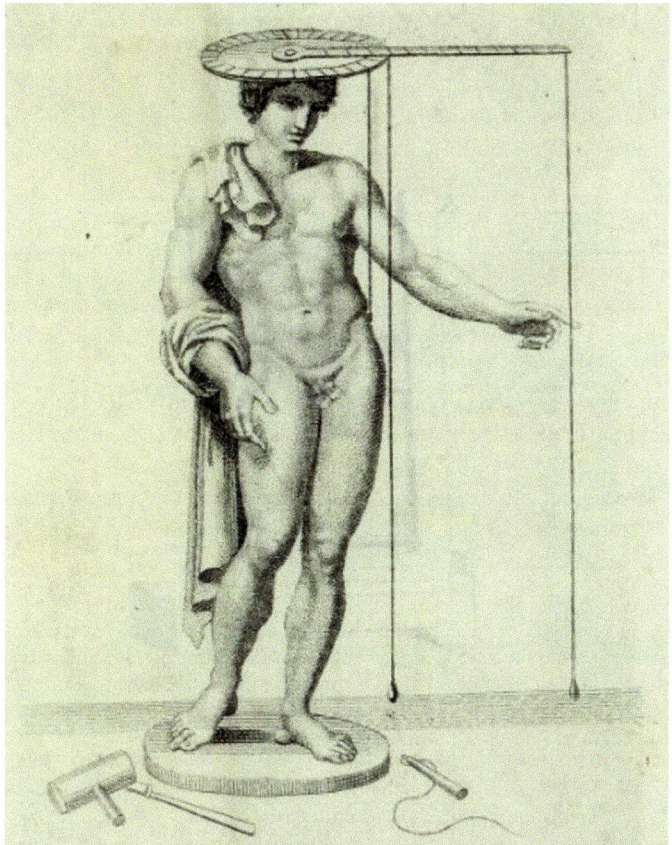

Fig. 7 Leon Battista Alberti, *De statua*, 1464

craftsmen [11]. This text marks the beginning of a dense and long epic story on the speculations of the scientific representation of the visual world.

9 The Geometric Order of Beauty

In the history of artistic literature, the German painter Albrecht Dürer occupies a prominent position, serving as a reference and inexcusable appointment for many other contributions after him. Dürer can also be considered as the major reference on the theoretical studies on proportions in the human body, through his written work.

In addition to the treatise on the proportions of the human body, the one related to geometry *Underweisung der Messung* was conceived, according to Dürer himself, to put in the hands of German painters, who, until that time (and in spite of their ingenuity and dexterity acquired by the practice in painting), had not been able to reach maturity, because they did not have a solid knowledge on the basis of painting: geometry, without which "no one can become or be a perfect artist" and that will be useful, not only for painters, "but also for goldsmiths, sculptors, stonemasons and, in one word, for all those who use a compass, a ruler and measurements".[9]

In his accurate words, the historian Panofsky pointed out that

> Dürer was the first artist who, trained in the late medieval workshops of the North, succumbed to the spell of the theory of art that had been created in Italy. It is in his development as an art theoretician that we can study in vitro, so to speak, the transition from a convenient code of instructions to a body of systematic and formulated knowledge. And it is in his contributions to that body of knowledge, written and printed, that we can witness the birth of German scientific prose [12].

The German painter is the strongest promoter of the study of the theory of proportions in the human body; a line of research will reach the contributions of Le Corbusier in the twentieth century with his *Modulor* (Fig. 8).

The works of Dürer can be considered, to a large extent, as precedents of modern scientific anthropometry. In fact, he did not want to pay attention to those who tried to improve nature by inventing artificial proportions:

> If the best parts chosen among many well-formed men come together in a single figure, the result will be worthy of consideration. But there are some with different opinions that judge how men should be made [...] I maintain that the perfection of form and beauty is contained in the totality of humanity. This is the model I will follow, from which perfection can emerge, and not from the one who invents a new body of such proportions that cannot be found among men [13].

Dürer's line of research, applied to the geometric conceptualization of natural shapes, will be consolidated in the nineteenth century thanks to Goethe's work about the term "morphology", introduced by the German writer around the year 1827, to establish analogies between organic and inorganic structures.

Works on the laws governing the relations of natural forms with mathematics and artistic creations would reach its maximum development with the publication of the works by Sir Theodore Andrea Cook and D'Arcy Wentworth Thompson on natural morphology [14] (Fig. 9).

The publication of descriptive geometry lessons by Gaspard Monge has been recognized as the milestone that marks the culmination of what is strictly considered as the science of graphic representation. Up to this event, many technical contributions from different trades were added, as well as those from artists and scientists.

Beyond the scientific importance that can be granted to it, descriptive geometry was imposed as a fundamental piece in teaching systems for the role assigned from the founding, in France, of the Normal School and the Polytechnic School. The French model would serve as reference for many teaching institutions in Europe and America.

---

[9]A Spanish translation of Dürer's geometry treat exists: *Instituciones de Geometría*. México: Universidad Nacional Autónoma, 1987.

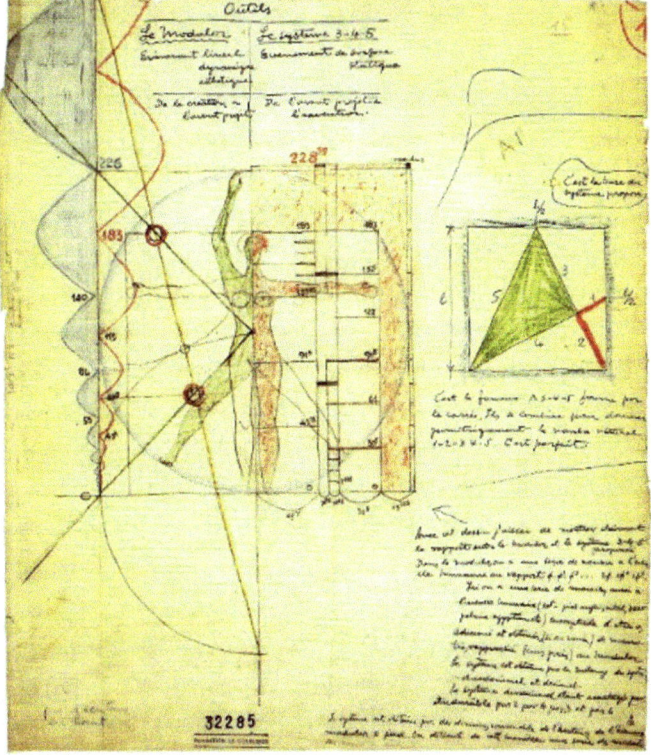

Fig. 8 Le Corbusier, *Modulor*, 1945

As a consequence derived from the French Revolution, the concept of National Education was applied in general education programs developed since then in many European states, according to the French model. This will be the definitive impulse to justify and demand the widespread existence of textbooks and, among them, obviously, those about geometry.

Five years before the publishing of the first edition of Gaspard Monge's *Lessons* had passed, and the first translation in the Imprenta Real de Madrid was published in Spanish, as a textbook named *Geometría descriptiva. Lecciones dadas en las Escuelas Normales en el año tercero de la República, por Gaspar Monge, del Instituto Nacional. Traducidas al castellano para el uso de la Inspección General de Camino.*

Like other enlightened projects, the first public institution of higher technical education, founded in the year 1802, after the parenthesis of the Spanish War of Independence, was aborted by the Fernandine reaction, that dissolved the Corps of Civil Engineers and the School; surprisingly, a chair of bullfighting was created in compensation [15].

This event supposes a transcendental fact in the history of technical lessons in Spain, since there are reasons to think that the application of the new geometric

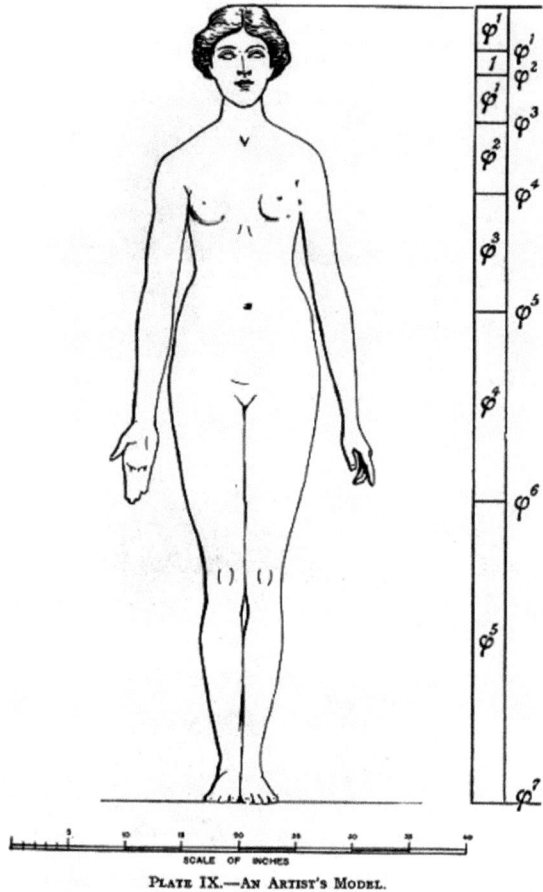

Fig. 9 Theodore Andrea Cook, *The Curves of Life*, 1914

Plate IX.—An Artist's Model.

theories had some curious and indirect consequences in the Spanish sociopolitical events.[10] In relation to this fact, it has been said that, from "a certain logic, following a specific human reason, 'mathematics' or 'geometry' could be subversive disciplines in the determined situation of the stately society" [16].

On the revolutionary ideology that Monge professed, the biographical data largely confirms this type of assertions, although it should not be understood that, thanks to it, anyone interested in the advancements of the mathematical sciences had necessarily to become a radical supporter of liberal policies. Despite this, there are obvious coincidences between students of geometry and social reformers, and this, in itself, is a significant fact.

---

[10] A facsimile edition of Monge's Descriptive Geometry exists, which contains a rigorous study by: Gentil Baldrich and Rabasa Díaz, "Sobre la Geometría descriptiva y su difusión en España", in: *Gaspard Monge. Geometría Descriptiva*. Madrid: Colegio de Ingenieros, Canales y Puertos, 1996.

The rational connotations of science in general and of geometry in particular were associated, in the first half of the nineteenth century, to political liberalism. This fact explains the failure of a teaching proposal of descriptive geometry, following the French example, published in a book in 1845, destined, according to the author, to the

> Colleges of secondary education where all the young people of a province concur to receive an education common to all careers; and because normal schools are understood in the same way, to where it is necessary to lavish a teaching, with a certain character of universality between certain boundaries.

Curiously, in this work's prologue "To the Queen" the author declared something suspicious of being accused as a liberal "the desire to contribute to the advancement of general education", adding that his decision when publishing the text, "As none has yet been written, and therefore its exclusive application is to be a part of the teaching of philosophy".[11]

10 Geometry in the Royal Academies

Prior to the existence of master education plans, in the Academies of Fine Arts of the last third of the eighteenth century, the low cultural level required for students to access them contributed to filling the gaps and partially occupying the place of the elementary school. For this reason, some facts can be explained, such as the inclusion in the curriculum of the Royal Academy of Fine Arts of San Fernando in 1821. Of a preliminary instruction establishing that, before starting to learn drawing, it was essential to have acquired enough knowledge about arithmetic to solve problems and then know how to represent different geometrical figures taking into account their mathematical foundation on a blackboard and paper [17].

For this particular purpose, some elementary level works expressly aimed at young Fine Arts students throughout Europe already existed. In Spain, we can highlight the *Instituciones de Geometría práctica para uso de los jóvenes artistas* [18] written by Bails and the *Principios de Geometría Descriptiva para los alumnos de pintura y escultura* [19] written by Laviña.

The scarcity of texts in Spanish explains why some art texts, despite being written in other languages, were translated and used for the quality of their images; this is the case of Gerard Audran's book, *Les proportions du corps humain*, published for the first time in Paris in 1683, and followed by reprints and translations such as the one made for the San Fernando Academy by Spanish Gerónimo Antonio Gil in 1780 [20] (Fig. 10).

---

[11]Even if the work's title is misleading, since the actual conception about delineation is associated with a profession or trade, it refers here to an abstract question related to mathematical nature when it speaks about "lines related". Thus, this work must be understood as a "lines' treatise". Gómez Santa María, Agustín, *Tratado de Delineación*. Madrid, Pedro Mora, 1845.

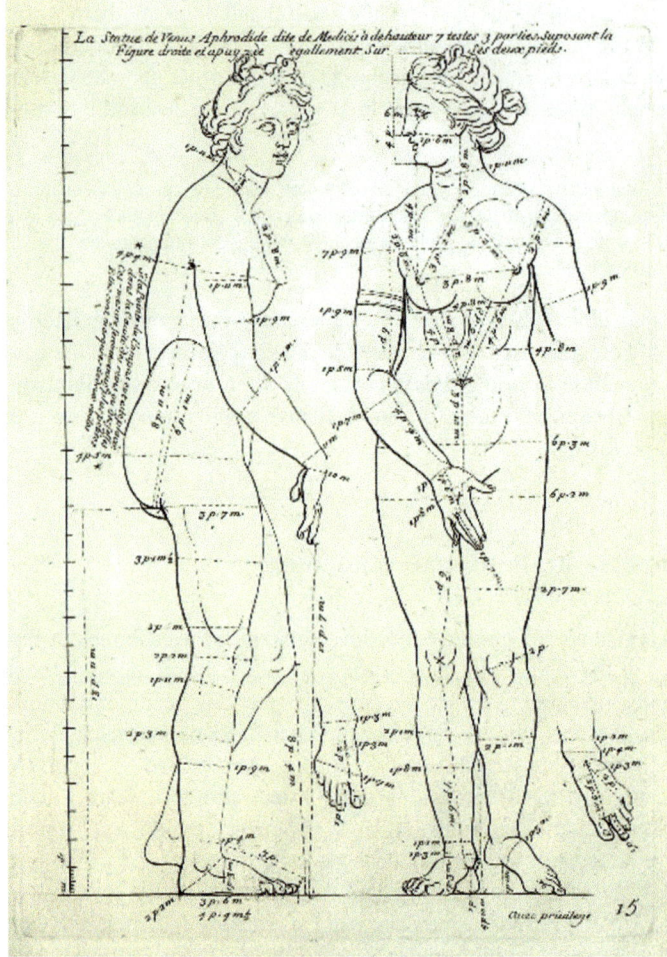

Fig. 10 Gérard Audran, *Les Proportions du corps Humain...*, 1693

11 The Current Landscape

From the perspective of mathematicians, geometry or theoretical geometries and their connection with practical applications of drawing are dispensable. In this sense, the words of a well-known mathematician can be remembered: "Drawing is forbidden for us, but nothing prevents us from continuing to use the language of geometry" [21].

On the contrary, practical geometry is associated with technical drawing, design, and multiple professional applications. There is a growing tendency to dispense those geometry subjects that only have a remote application for practice; it is the case of many topics related to projective geometry or a good part of the mathematical theory of axonometry.

Computer graphics created with CAD programs have forced the substitution of descriptive geometry by a constructive or generative geometry of forms; likewise, geometric drawing has been affected by what is already known as "electronic delineation." In this sense, computer-assisted drawing should be understood more as a conceptual instrument than as a material resource. The need of a certain balance between the intuitive methods based on the graphic control of figures and traditional abstract reasoning is maintained; undeniably, with graphic computing, the resources and possibilities for the visualization of figures have been enhanced.

References

1. Bédat, C.: La Real Academia de Bellas Artes de San Fernando (1744–1808), p. 218. Fundación Universitaria española, Madrid (1989)
2. de Honnecourt, V.: *Cuaderno*. Akal, lám. 36, Madrid, p. 134 (1986)
3. Euclides.: *Elementos*. Introduction by Luis Vega. Translation and notes by María Luisa Puertas Castaños. Gredos, Madrid (1991–1996)
4. Hilbert, D.: *Fundamentos de la Geometría*. Introduction by José Manuel Sáchez Ron. Consejo Superior de Investigaciones Científicas, Madrid (1991)
5. Brunschvicg.: *Les étapes de la Philosophie mathématique*, París, 1947. Quoted in: VEGA, Luís. "Introduction" to los *Elementos* de Euclides, p. 47. Gredos, tome 155, Madrid (1991)
6. Scolari, M.: La prospettiva gesuita in China. Casabella **507**, 49 (1984)
7. La Perspectiva y Especularia de Euclides. Traducidas en vulgar Castellano, y dirigidas a la S. C. R. M. Del Rey don Phelippe nuestro Señor. Por Pedro Ambrosio Onderiz su criado, Alonso Gómez's widow, Madrid (1585)
8. Geometría y Traça para el oficio de Sastres. Fernando Díaz, Sevilla (1588)
9. Garriga, J.: La *intersegazione* de Leon Battista Alberti (I). In: *D'Art* 20, 1994 Revista del Departamento de Historia del Arte, pp. 13–14. Universidad de Barcelona
10. Rejón de Silva, D.A.: El Tratado de la Pintura por Leonardo de Vinci y los Tres Libros que sobre el mismo Arte escribió Leon Bautista Alberti. Traducidos e ilustrados con algunas notas por Don D. A. R. De S, p. 197. Imprenta Real, Madrid (1784)
11. Pélerin, J. (Viator): De artificiali perspectiva. Toul (1505)
12. Panofsky, E.: Vida y arte de Alberto Durero, p. 255. Alianza, Madrid (1982)
13. Sholfield, P.H.: Teoría de la proporción en arquitectura, p. 59. Labor, Barcelona (1971)
14. Cook, T. A.: The curves of life. Nueva York: Dover. (original edition, London, 1914). Thompson, D'Arcy Wentworth, *On Growth and Form*. Cambridge: Univ. Press (1979). (1917's original edition)
15. de Orduña, C.: Memorias de la Escuela de Caminos. Madrid (1925)
16. Clavero, B.: Razón científica y revolución burguesa. In: El científico español ante su historia. La Ciencia en España entre 1750–1850, p. 229. Diputación, Madrid (1980)
17. Matilla, J.M.: Las disciplinas en la formación del artista. In: La formación del artista, de Leonardo a Picasso. Real Academia De BB. AA. De San Fernando, Madrid (1989)
18. Bails, B.: Instituciones de Geometría práctica para uso de los jóvenes artistas. Vda de Ibarra, Madrid (1795)
19. Laviña Blasco, D.M.: Principios de Geometría Descriptiva para los alumnos de pintura y escultura. a. Vicente, Madrid (1859)
20. Gil, G.A.: Las proporciones del cuerpo humano, medidas por las más bellas estatuas de la antigüedad/ que ha copiado de las que publicó Gerardo Audran. Joachim Ibarra, Madrid (1780)

21. Boll, M.: Historia de las matemáticas, p. 119. Diana, México (1970)

Lino Cabezas Gelabert is a Doctorate in Fine Arts and Full Professor in Drawing from Barcelona's University. Formerly, he was Professor at E. T. S in Catalunya's Architecture Polytechnic University. Lino Cabezas is a specialist in perspective, technical drawing and geometry. Author and co-author of several books, including the ones published by Ed. Cátedra de Madrid: *Las lecciones del dibujo*; *El manual de dibujo*; *Máquinas y herramientas de dibujo*; *El Dibujo como invención*; *Idear, construir y dibujar*; *Los nombres del dibujo*; *La representación de la representación*; *Dibujo y construcción de la realidade*; *Dibujo y territorio*. Lino Cabezas Gelabert was one of the keynote speakers of the international conference Geometrias'17.

The vaults of Arronches Nossa Senhora da Assunção and Misericórdia churches. Geometric and constructive comparison with the nave and refectory vaults of Jerónimos Monastery

Soraya M. Genin

Abstract The comparison between the vaults of the Arronches Nossa Senhora da Assunção church and the Jerónimos monastery church was first performed by Mário Tavares Chicó [1]. He noticed how the naves and side aisles were covered by a single curved shape, unique in Europe. Mendes Atanázio attributed both authorships to one architect: João de Castilho (c.1470–1552). The architectural analysis of both vaults concluded that there were several geometrical similarities, particularly in relation to their elevation and their gothic proportions. Arronches Misericórdia church was built in the same period as the Assunção church. It is covered by a *liernes* vault, similar to the refectory vault of the Jerónimos monastery. The geometric analysis of both vaults shows a similar design, in plan and in elevation. The geometric analysis of the vaults was obtained from architectural surveys. Design and construction hypothesis presented were developed using the methods of Hernán Ruiz (1500–1569) and Rodrigo Gil de Hontañón (1500–1577). They were both Spanish architects, contemporaries of João de Castilho.

Keywords Manueline ribbed vaults · Geometric and constructive analysis · Late Gothic Architecture

1 Introduction

The construction of Arronches Nossa Senhora da Assunção church dates from the beginning of the sixteenth century. According to Luis Keil, it was built on a primitive construction from 1236 to 1242:

This paper is based on the doctoral thesis "Voûtes à nervures Manuélines. Le caractère innovant de João de Castilho," published in 2014 [2].

S. M. Genin (✉)
School of Technology and Architecture, ISCTE - Instituto Universitário de Lisboa, Lisbon, Portugal
e-mail: soraya.genin@iscte-iul.pt

© Springer Nature Switzerland AG 2020
V. Viana et al. (eds.), *Thinking, Drawing, Modelling*,
Springer Proceedings in Mathematics & Statistics 326,
https://doi.org/10.1007/978-3-030-46804-0_5

There are still faint traces on which the temple may have been raised, although tradition claims the site of the first church is on the left of the present church. What seems certain is that the tower was erected on the foundations of the twelfth century temple. [3].

Mário Tavares Chicó compared the Assunção church's vault to the vaults of the churches of Freixo de Espada-a-Cinta and of Jerónimos monastery. He noticed a common characteristic that was a novelty in Europe: The merging of the nave and side aisles, more achieved than in the late German gothic [1].

No documentary source refers to the authorship of Arronches church. Mendes Atanázio, through architectural analogy, defends the attribution of the three vaults to one architect, João de Castilho (c.1470–1552), documented as master of the Jerónimos monastery since 1517. Therefore, he deduces that Freixo de Espada-a-Cinta would be the first of the three vaults to be built, followed by the vault of Arronches and, finally, by the vault of Jerónimos, the latter being the most complex and perfectly shaped:

Arronches mother church planimetry is quite similar to the one of the Belém church, though shorter and showing no star-shaped ramifications in its vault, and having columns instead of octagonal pillars; from one support to another, in the transversal direction, there are ribs that are also curved at three central bosses, in the middle of the bays. The differences of structural scheme between Arronches and Belém leads us to conclude that there was an architectural consciousness trying new solutions in order to unify the space, which is creative and dynamic. Viewed from the extrados, Arronches vault gives the impression of being composed of successive parts of vault, covering each one bay, a question that deserves an attentive study. [4]

This typology is unique in Europe, as confirmed by the analysis of approximately 1000 European vaults, achieved through bibliography research and visits, including France, England, Germany and Central Europe, Spain and Portugal [2].

In Portugal, in addition to the three vaults mentioned in the bibliography, two similar ones were identified, presenting a curved shape across the nave and side aisles: in Viseu's cathedral and in Torre de Moncorvo's church. Considering a formal evolution, aspired by João de Castilho, the vault of Viseu's cathedral would have been the first one, since the curved form achieved there is less perfect and does not fit in an arc of circumference, as in the other vaults [5].

This paper presents more details about the geometric and constructive analysis of the nave vault of the mother church Nossa Senhora da Assunção. It also presents the analysis of the nave vault of Arronches Misericórdia church, located next to the mother church, as well as the refectory vault of Jerónimos monastery, both similar in typology and form. As both churches were built in the first half of the sixteenth century, it is not surprising that they were the work of the same architect, and even derived from the same work contract. The geometric and constructive analysis of the simplest vaults, the cross-ribbed vault, is presented first, followed by the *liernes'* vaults.

2 Cross-Ribbed Vaults—Geometry and Construction

Since the beginning of the construction of cross-ribbed vaults, the vaults shape has always been dependent on the shape of the ribs. Diagonal ribs were initially semi-circular and later oval, in order to lower the vaults. Depending on the constructive method used, the *rampante* (axis of the vault) was horizontal or curved. The curved *rampante* is characteristic of the vaults in Spain, João de Castilho's country of origin.

Ribs multiplication using *tiercerons* (ribs going from the supports to secondary bosses) and *liernes* (ribs between bosses) allowed to shape the vault and to perfect the space unification. It was also intended to facilitate the construction, as the webs of the vault being supported by the ribs, without the need for more intermediate supports. The rib was the main element, the skeleton of the vault, which allowed to create the form and to construct the vault.

Rib standardization is documented since the thirteenth century. Its purpose was to simplify the construction of the centrings and the carving of the ribs. Viollet-le-Duc refers to Villard de Honnecourt's drawing that illustrates the way to use the same semicircle for all the ribs (Fig. 1a) [6].

This principle of standardization was confirmed in 43 of the analysed vaults. The geometry was studied, from an architectural survey of the ribs, using a distance metre.

The design and construction of ribbed vaults are documented in two sources of the sixteenth century: a drawing by Hernán Ruiz (1500–1569) and the indications of Rodrigo Gil de Hontañón (1500–1577) described in the Simon Garcia compendium [7].

In the following, we present Rodrigo Gil's indications for the construction of a cross vault: at first, the work platform is placed, on which the plan is drawn, including the location of all the bosses; the diagonal centring is assembled; the location of the

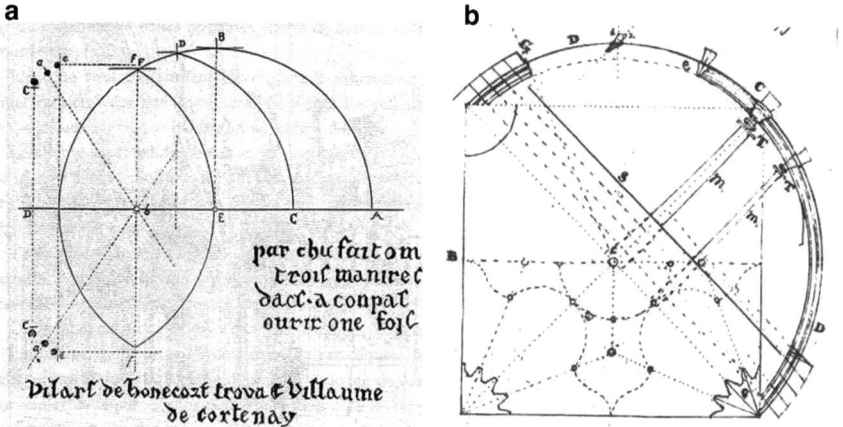

Fig. 1 **a** Interpretation of Villard de Honnecourt's design, according to Viollet-le-Duc [6]; **b** drawing of Rodrigo Gil de Hontañon for the construction of a ribbed vault (Simon Garcia) [7]. Both drawings document the rib standardization during the thirteenth and sixteenth centuries

central boss is marked on the centring with a plumb line aligned to its location on the platform; from this boss, the *rampante* is assembled (the same diagonal's curve is used, so it is a domed shape); the bosses are located on the *rampante* with a plumb line; struts are placed to support the bosses; then the centrings of the remaining ribs are placed in these struts. The bosses are always located with a plumb line, going down from the centring, the diagonal or the *rampante*, according to the plan drawn on the platform (Fig. 1b).

Rabasa Díaz presents and interprets a drawing by Hernán Ruiz (Fig. 2), which shows the following order: first, the semicircle arc of the diagonal *AC* is drawn; from the central boss is drawn the *rampante BD*, whose length is taken from the plan and whose radius is equal to the diagonal; bosses *D* and *E* are located in this arc, the heights of the formers and the *tiercerons* are measured; arcs *AC* and *AF* are drawn with the centre at the level of the diagonal, in front view, with the length taken from the plan. Rabasa Díaz comments on Ruiz's various attempts to find the centre of the circumference arcs [8].

It is interesting to notice that the order of construction, mentioned by Rodrigo Gil, is the same as the order indicated by Hernán Ruiz in his drawing: first the diagonal and the central boss, then the *rampante* and the bosses on the *rampante* and, finally, the remaining ribs. The use of this method, by João de Castilho, for the construction of diverse types of ribbed vault has been proven by the author [9].

a   **b**

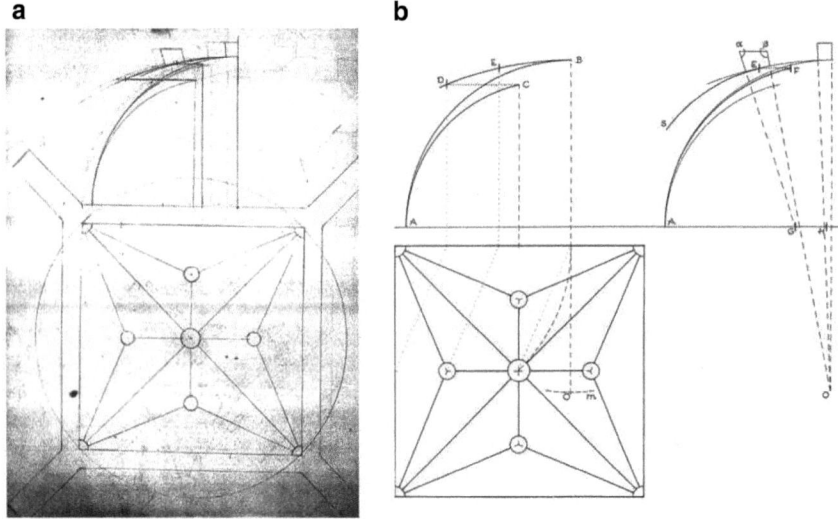

Fig. 2 **a** Drawing by Hernán Ruiz: plan and elevation of a *tiercerons* vault; **b** Drawing Interpretation, by Rabasa Díaz [8]

3 Arronches Assunção Church—Lateral Chapel Vault

The lateral chapel of Arronches Nossa Senhora da Assunção church presents a 5.60 m × 3.82 m floor plan and a height of 7.52 m. It is covered by a cross-ribbed vault (Fig. 3). For a better understanding of the analysis of the following vaults, this vault is presented first, since it shows a simpler typology.

3.1 *Design and Construction Hypothesis*

First, the plan is drawn with the desired dimensions and then the circumference of radius *M1*, which is the standard arc (*p*). For the elevation, the arc of the diagonal in semicircle is drawn and the central boss 1 is located. From height 1 and the length

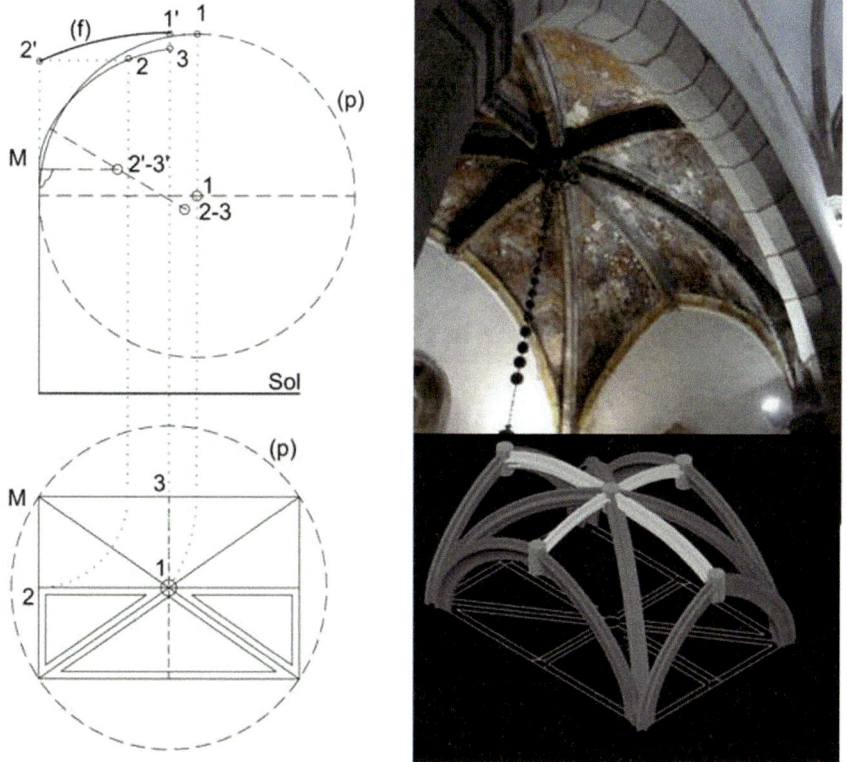

Fig. 3 Right chapel of Arronches Nossa Senhora da Assunção church. Hypothesis for the design of the plan and elevation of the ribs. The circumference that circumscribes the plan is the standard arc (*p*), common to all the ribs. The *rampante* (*f*) is inscribed in an arc of radius equal to twice the radius of (*p*). The three-dimensional drawing emphasizes the two curved *rampante*s of the vault [2]

of the *rampante* taken from the plan, it is located point 1′. From 1′ the arc of the *rampante* (*f*) is drawn, with a radius equal to twice the radius of (*p*). Point 2′ is located at the intersection with the vertical of the wall. Once these two arcs are represented, formers arcs *M2* and *M3* are drawn. As for the latter arc, in order to draw the elevation of *M2*, its elevation is first measured from point 2′ located on the *rampante* and its distance from the plan, and point 2 is located[1]; From 2, the longer arc is drawn with the radius of (p) and the minor arc tangent to (p) vertically to the wall. Arc *M3* is identical to *M2*.

The arc of the diagonal is the main arc (*p*), a semicircle, which is used for all the ribs. The *rampante* is inscribed in an arc of circumference (*f*) with a radius equal to the length of the diagonal in plan, that is twice the radius of (*p*), in an identical drawing to that of Hernán Ruiz.

The shape of the *rampante* is visually broken, because this curve is constructed with two segments identical to (*p*). It means that there is a preliminary design that defines the arc (*f*), but the construction is simplified using a single arc (*p*) for the carving of the ribs.

4 The Nave Vault of Nossa Senhora Da Assunção Church

4.1 Comparison with the Vault of Jerónimos Monastery Church

Arronches Assunção church is composed of five bays and presents a plan of 30.70 m × 15.30 m. The vault rests on 1.00-m-thick walls and thin columns with 0.73 m in diameter. On the outside, the buttresses are aligned with the columns and positioned diagonally in the wall-corners. The bays are composed of transverse and diagonal ribs. There are no longitudinal ribs, and the central nave and the side aisles merging into a single space, favouring the transversal and diagonal reading of the space. The transverse ridge-line connects the central bosses of the nave and of the side aisles, with 11.60 m and 11.40 m height. The diagonal ribs are defined by semicircles (Fig. 4a).

Jerónimos monastery church presents five bays and a plan measuring 50.80 m × 22.40 m. The pillars are 17.00 m high and only 1.00 m thick, and their profile is octagonal. No orthogonal ribs limit the bays, and no diagonals can be seen between the supports. Instead, the traditional ribs are replaced by opposing triangles pairs, formed by *tiercerons* and *liernes*. The curved ribs draw figures—octagons, squares—which give more amplitude to the summit of the vault. The curved shape is not just the transverse ridge line. This arrangement allows the vault to gain a visually enlarged, barrel-shaped surface at the summit. The height of the central bosses (24.00 m in the

---

[1]This method, that consists in taking the dimension from the plan and the height from the *rampante* in order to trace the elevation, is used by Hernán Ruiz and other contemporary architects, as shown in the drawings of the time.

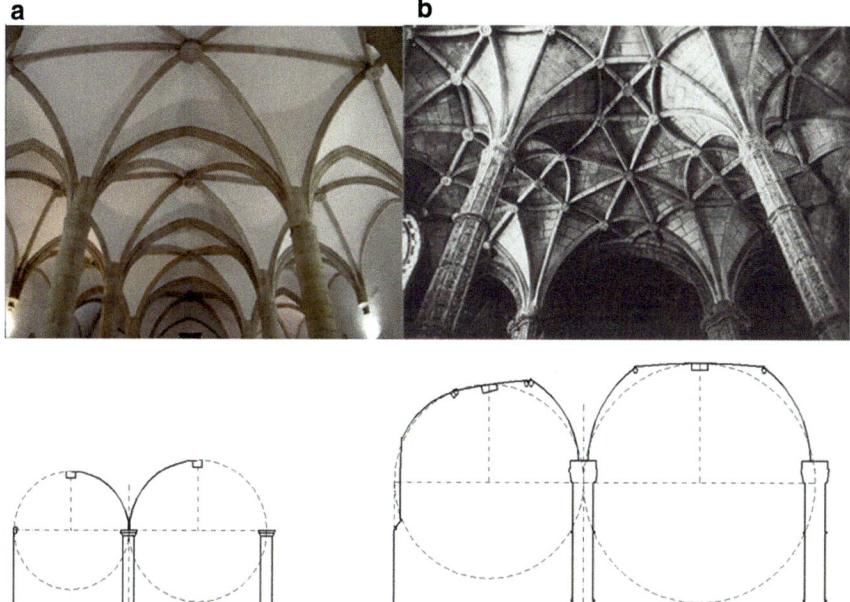

Fig. 4 **a** Arronches church and Jerónimos monastery nave's vault geometry. **b** The height of the central bosses of the side aisles and the naves corresponds to the height of diagonals in semicircle [2]

nave and 23.20 m in the side aisles) corresponds to the radius of a semicircle on the diagonal, as in Arronches, although it is not materialized by any rib (Fig. 4b).

Both vaults differ in plan, but present a curved section responding to identical design and geometric principles. The height of the central bosses is determined by semicircles on the diagonals. Theoretically, three points (the central bosses) allow tracing the transverse *rampante* arch. The shape is based on the habitual geometry of cross-ribbed vaults, maintaining the traditional proportions, therefore ensuring stability.

4.2 Plan

- The vaults are drawn in plan; the central bay is approximately square and the lateral one presents a 3:4 proportion.
- The axes and the diagonals are drawn, and then their bosses are located at the intersections.

4.3   Elevation

- Diagonal: in elevation, the diagonals are semicircles (*p*) whose radii are taken from the plan. The keystones 1 and 2 are located on the vertical axes that pass through the centre of the diagonals.
- *Rampante*: the curve of the transversal *rampante* (*f*) is drawn from bosses 1 and 2, and from a third point symmetrical to 2 (keystone of the right-side aisle, not shown in the drawing).
- Transversals and formers arcs: the circumferences are drawn with radii *p1* and *p2*. Arc *P4* is coincident with *P1*. Arc *P5* is located lower than *P2*, since boss 5 is also lower than boss 2. *M3* and *M5* are two-centred arcs whose long arc is identical to (*p2*), while the short arc is tangent to the wall and to (*p2*). Two methods of constructing ribs from the same standard arc can be observed: by vertical translation of the arc until reaching the desired boss height or, when the arc does not fit in the space between the *tas-de-charge* and the boss, by using a shorter arc near the haunches, tangent to the main arch and the wall[2] (Fig. 5).

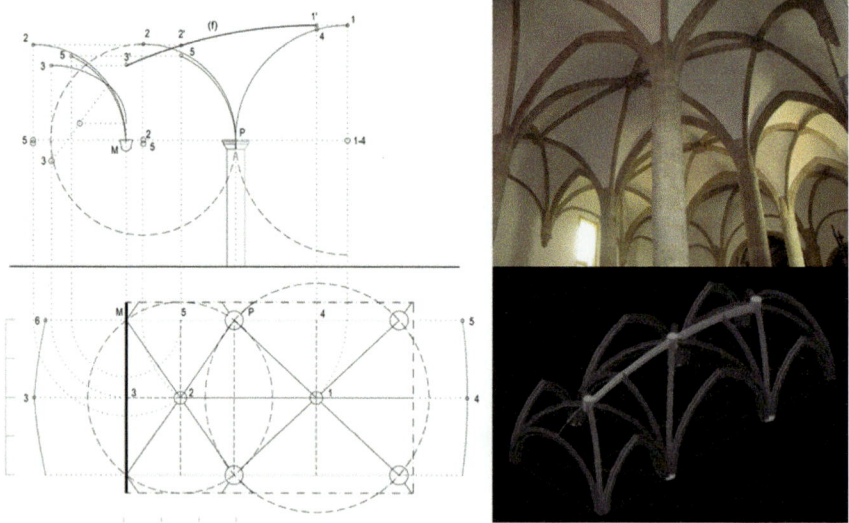

Fig. 5   Nave vault of Nossa Senhora da Assunção church. Hypothesis for the design of the plan and the elevation of the ribs. The semicircle diagonals are the standard arcs (*p1*) and (*p2*) that define all the ribs. The three-dimensional design of the vault emphasizes the curved line of the transverse *rampante* that spans the nave and side aisles, unifying them visually [2]

---

[2]These two methods were frequently used during the Gothic period and are confirmed in the Manueline vaults, as it could be verified in the case of the 43 vaults analysed [2].

4.4 Construction

The diagonals, semicircular ribs, are the standard arcs *p1* and *p2*, which define all the ribs. The ribs present different profiles: the transverse ribs are the thickest ones (0.37 m), followed by the diagonals (0.30 m) and finally the ribs of the *rampante* (0.25 m).

- Plan: the full-scale plan is drawn on the work platform in which the centrings are placed.
- Diagonals: the *centrings* of the diagonals are first placed; the bosses' location on the *centring* is found, using a plumb line aligned according to its position on the work platform.
- *Rampante*: a *centring* is placed for the transverse *rampante* between the three central keystones. The boss of the formers is set in its position.
- Transverse ribs: the *centrings* of the transverse ribs are placed. The *centrings* of the longitudinal *rampantes* are placed between their tops and the central bosses.

After placing the *centrings*, the bosses of the diagonals and then the *voussoirs* of the ribs are placed from the *tas-de-charges* to the summit of the vault. The panels between the ribs are filled with brick masonry. The haunches are filled with masonry.

5   The Nave Vault of Arronches Misericórdia Church

The nave of Arronches Misericórdia church is covered by a *liernes* vault, shaped as a four-pointed star. The main structure consists of cross ribs and *tiercerons* between the corbels (Fig. 6).

The plan dimensions of the nave are 9.80 m × 6.70 m for a height of 8.70 m. All the ribs are defined by the two circle arcs that compose the two-centred arc of the diagonal. The diagonal is lower than the other ribs and 0.20 m higher than the corbel. The centre of the short circle arc is located at 1:4 of the length of the diagonal. The transverse *rampante* is curved, and its radius is equal to the length of the diagonal in plan, and twice the previous arc (the same relation was found in the design of the chapel, as the one of Hernán Ruiz). The ribs are identical in profile and thickness (0.30 m).

5.1 Plan

- The diagonals and both *rampantes* are drawn from a rectangular plan. Bosses 1, 2 and 3 are located.
- The bisector *M4* and line *2-M1* are drawn. At the intersection of these two lines, boss 4 is located. Boss 5 is located at an identical distance from *M* as boss 3.

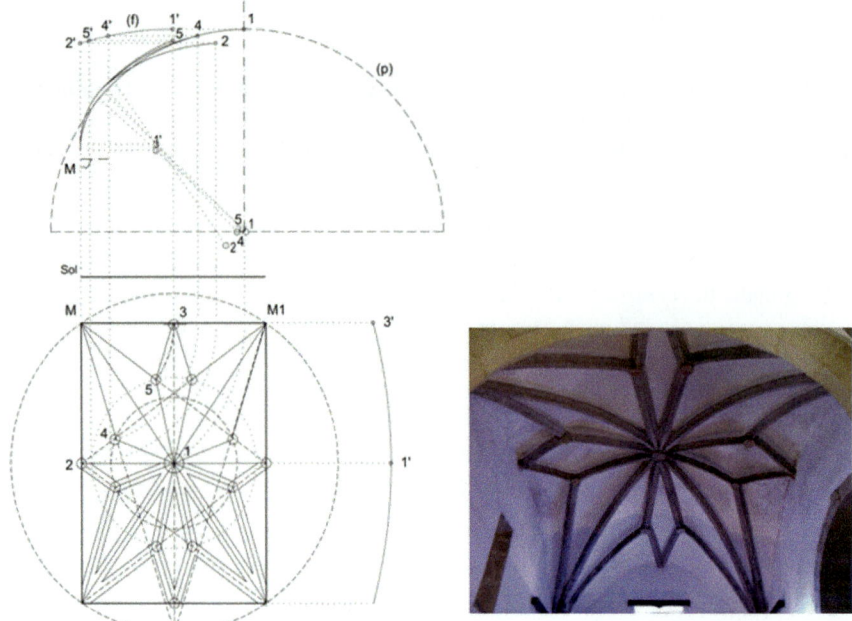

Fig. 6 Nave vault of Arronches Misericórdia church. Hypothesis for the design of the plan and the elevation of the ribs. All ribs are drawn with the two semicircle arcs that define the diagonal. The circumference that circumscribes the plan corresponds to the long arc (*p*). The *rampante* (*f*) is drawn with a radius twice the radius of (*p*) [2]

5.2 Elevation

- Diagonal: the diagonal is drawn as follows: distance *M1* is taken from the plan and the vertical axis of the vault can be drawn; the arc of the diagonal is outlined with a two-centred arc; the longest arc is identical to (*p*) and centred on the pavement; point 1 is located at the intersection with the vertical axis.
- *Rampante*: the *rampante* can be drawn from 1. Its radius is equal to the length of the diagonal in plan. In this line, bosses 4 and 5 are positioned. Their distances from boss 1 are taken from the plan.
- *Tiercerons* and formers: these ribs are drawn with the same arcs as the diagonal, but their centres are located higher. In order to reach the required boss (2, 4 or 5), the long arc rotates around the centre of the smaller, until their height, previously determined by the *rampante* line, is found.

5.3 Construction

The centrings are constructed, and the ribs are carved according to the standard arc (*p*). The construction follows the same order described for the design.

- Plan: the plan is drawn on the work platform, showing the location of all bosses.
- Diagonal: the centrings for the diagonals are placed; the central boss 1 is located with a plumb line directed to its position on the work platform.
- *Rampante*: from boss 1, the centring for the transverse *rampante* is placed and the bosses are located with a plumb line, according to their positions on the work platform.
- *Tiercerons*: following the heights of the *rampante*, the struts of bosses 4 and 5 are placed in their marked position on the work platform. The centrings of the *tiercerons* are placed between the struts and the *tas-de-charge*. The process is the same for all four sides.
- *Liernes*: As soon as the *tiercerons* centrings are in place, the *liernes* centrings are placed between the bosses.
- After completion of the centring work, the bosses are placed on the top of the struts, and the *voussoirs* and webs of the vault are built starting from the *tas-de-charge* to the central keystone. The haunch area is filled with masonry. Finally, the vault is covered with a thin layer of masonry, forming the extrados.

6 Jerónimos Refectory Vault

The refectory of Jerónimos monastery presents five bays, covered by *liernes* vaults shaped as four-pointed stars. The vault is similar to the previous one, but presents no cross ribs. It is only composed of *tiercerons* shaped as four-pointed star. In the plan, an eight-pointed star is observed (Fig. 7).

The bay has dimensions of 9.00 m × 7.68 m in plan. The height at the central keystone is 7.85 m; about twice the height of the corbel, this is at a height of 3.80 m. The arches are lowered and the impost is located 0.95 m above the corbel.

Geometric analysis reveals that a standard two-centred arc was used to construct all the ribs. The long arc has a radius equal to the circumference that circumscribes the plan (*p*). With the exception of the former, the centre of the shorter arc is at the same level for all the ribs. The centre of the long arc rotates around the previous centre, so that the curve meets the boss height, which has an identical design to that of Misericórdia church in Arronches.[3] The two *rampantes* correspond to an arc of circumference whose radius is equal to half the length of the diagonal of the refectory. The summit of the vault has, therefore, a domed shape.

---

[3]This same method was used for other vaults of Jerónimos monastery, namely in the north chapel of the church, the ground floor of the cloister and the sacristy. It is frequently found in English vaults [2].

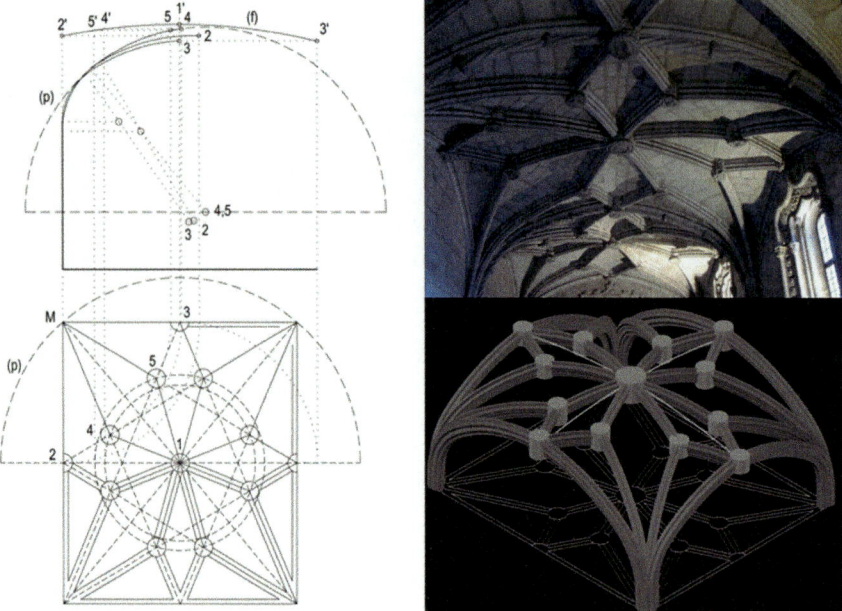

Fig. 7 Refectory vault—Jerónimos monastery, Lisbon. Hypothesis for the design of the plan and the elevation of the ribs. The ribs are two-centred. The long arc is identical to (p), except for the former. The standard arc (p) rotates around the centre of the short arc to reach each boss. The boss heights are determined by the curve of the *rampante* (f). The *rampantes* corresponds to an arc of circumference, whose radius is equal to half of the diagonal of the refectory. The three-dimensional illustration shows the curved *rampante*, which determines the bosses' heights and produces a domed shape at the summit of the vault [2]

6.1 Plan

- First, the diagonals and the axes of the vault are drawn. The central boss 1 and the bosses of the formers and transverse ribs, 2 and 3 are located.
- A straight line is drawn between 2 and 3 and the opposing supports. Their intersections define the bosses of the *tiercerons*, 4 and 5. From these bosses, the *tiercerons* and *liernes* are drawn.

6.2 Elevation

- *Rampante*: the arc of the *rampante* (f) is drawn, from the central boss, with pre-defined height. 1', 2' and 3' are located on the *rampante*, with the distances measured from the plan.

- Formers: bosses 2 and 3 are located with distances *M2* and *M3* taken from the plan and their heights 2′ and 3′ taken from (*f*). The centres of the two-centred arc are found, knowing that the long arc (*p*) has a radius equal to *M1*. The short arc of the former *M3* is raised in relation to the remaining ribs.
- *Tiercerons*: bosses 4 and 5 are located from the heights of 4′ and 5′ measured from the *rampante*. Once their location is found in the elevation plan, the *tiercerons* are drawn with the same arcs. The short arc is coincident, and the centre of the long arc rises and reaches bosses 4 and 5.

6.3 Construction

The centrings are placed, and the ribs are carved with curve (*p*). The ribs are carved with three different profiles. The transversal ribs are thicker (0.45 m) and show a rope-like decoration; the *tiercerons* and *liernes* are identical (0.40 m); the former's profile is different but with equal dimensions to the previous one (0.40 m).

- Plan: the plan is drawn on the work platform.
- *Rampante*: first the *rampante* centring (*f*) is placed, then boss 1 is located with the help of a plumb line, and the corresponding strut is placed on the work platform.
- *Tiercerons*: bosses 4 and 5 are located on the *rampante*, with distances 1–4 and 1–5 taken from the plan. The struts of bosses 4 and 5 are placed in their location on the working platform. Their heights are taken from the *rampante*. The centrings of the *tiercerons* are then placed.
- *Liernes*: the centrings of the *liernes*, between the bosses, have the same arc (*p*).

Once the centrings are completed, the bosses are positioned. Their shape is cylindrical, and they do not require the previous drawing. The *voussoirs* of the ribs and vault webs are placed. The arrangement in the vault is mostly concentric in close proximity to the central key. The summit of the vault has the shape of a dome.

The order of construction is identical to the previous one, except that it starts with the *rampante*, because there is no diagonal. The elevations of all bosses can be taken from the *rampante*.[4]

7 Conclusions

The results confirm the similarity between the nave vault of Nossa Senhora da Assunção church of Arronches and the nave vault of Jerónimos monastery church, particularly in elevation. The shape is defined by a curved *rampante* across the nave

---

[4]This is the process used by Castilho in Jerónimos monastery's nave for the bosses that are not located on the *rampante* [2].

and side aisles, which is the main element that establishes the elevation of all bosses. The heights of the central bosses proceed from a semicircle on the diagonal. The ribs are drawn with arcs of circumference or two-centred arcs, according to the bosses' elevation.

New architectural relationships can be found, namely between the nave vault of Misericórdia church in Arronches and the refectory vault of the Jerónimos monastery. These two vaults are lowered, and their diagonals are two-centred arcs. The shorter arc at *tas-de-charge* level is common to all ribs. The long arc rotates around the centre of the short arc in order to reach the desired height of the boss. The two *rampantes*, transverse and longitudinal, are identical and the summit is dome-shaped.

All the vaults show standardized ribs, method that was currently used during the gothic period to facilitate the construction of the centrings and the carving of the ribs.

The unification of the space and the continuity of the shape, achieved through the design and construction defined by the *rampante*, are characteristics of the João de Castilho's architecture, to whom the authorship of these vaults is attributed.

References

1. Chicó, M.T.: A arquitectura do Manuelino. Livros Horizonte, Lisboa (2005)
2. Genin, S.M.: Voûtes à nervures manuélines. Le caractère innovant de João de Castilho. Ph.D. thesis, Katholieke Universiteit Leuven (2014). https://lirias.kuleuven.be/handle/123456789/454010
3. Keil, L.: Inventário Artístico de Portugal - Distrito de Portalegre. Academia Nacional de Belas Artes, Lisboa (1943)
4. Atanázio, M.C.M.: A arte do manuelino. Presença, Lisbon (1984)
5. Genin, S.M., de Jonge, K., Palacios Gonzalo, J.: Portuguese vaulting systems at the Dawn of the early Modern period. Between tradition and innovation. In: Kurrer, K., Lorenz, W., Wetzk, V. (eds.) Third International Congress on Construction History. Cottbus, vol. 2, pp. 671–678
6. Viollet-Le-Duc, E.E.: Dictionnaire raisoné de l'architecture française du XIème au XVème siècle. Ed. De Nobele, Paris (1854–1868) (1967)
7. García, S.: Compendio de arquitectura y simetría de los templos, estudios introductorios de Antonio Bonet Correa y Carlos Chanfón Olmos. (facsímil.): Ed. Colegio Oficial de Arquitectos de Valladolid 1991 (1681)
8. Rabasa Díaz, E.: Forma y Construcción en piedra, de la cantería medieval a la estereotomía del siglo XIX. Ed. Akal, Madrid (2000)
9. Genin, S.M.: Form, design and construction of ribbed vaults. João de Castilho's innovations in the Jerónimos Monastery, Lisbon (1470–1552). Constr. Hist. Int. J. Constr. Hist. Soc. 33(1/2018):27–48 (2018). ISSN 0267-7768

Soraya Genin is graduated in Architecture at the Faculty of Architecture of Lisbon (1990), has a Master degree in Science in Architecture, Specialization Conservation of Historic Towns and Buildings (1995) and a Ph.D. in Engineering (2014) by the Faculty of Engineering at the KU Leuven.

Assistant Professor in ISCTE - Instituto Universitário de Lisboa and researcher at ISTAR-IUL, where she teaches Architectural Technology. Soraya Genin authored several studies and Architectural Conservation projects developed in her studio, established in 1999.

One of her main research interests is the conception and construction of ribbed vaults, mainly the geometric analysis on architectural design. Soraya M. Genin was one of the keynote speakers of the international conference Geometrias'17.

Perspective Transformations for Architectural Design

Cornelie Leopold

Abstract Geometric transformations enable to represent spatial objects in different ways dependent of the kind of transformation. The methods of representation had not been combined with the idea of transformation at the beginning of their development. An important step had been done in the intellectual history of perspective, by bringing together the mathematical concept of projections and, further on, transformations with the representational concept of perspective according to seeing and perceiving. The step to projective geometry gave the possibility to apply the perspective transformation on objects in space and to receive again spatial objects by the transformation. The consequence had been a systematically theoretical work out of relief perspective as spatial transformation, which started as theatre stage design. This comprehension enables applications in architectural design processes. The fruitful interlacing of practice in art and architecture with mathematical theory will be reflected and applied in architectural design as spatial transformation considering perception.

Keywords Transformation · Architectural design · Relief perspective

1 From Perspective to Scenography

The relationship between space and image, architectural spatial design and its perception is the starting point of our research. Perspective had been developed in order to produce images of space according to our seeing [1]. The important historical step for drawing perspectives to represent according to our seeing had been done by Alberti. The further developed perspective machines, for example by Albrecht Dürer, helped to translate the concept of perspective in practical guiding tools for the production of the perspective images, therefore, as a tool in art. The mathematical concept of the perspective transformation as collineation later brought the possibility to apply the transformation to the spatial object, in order to receive a transformed spatial object. The perspectival transformed spaces or objects started in the sixteenth century with

C. Leopold (✉)
TU Kaiserslautern, Kaiserslautern, Germany
e-mail: cornelie.leopold@architektur.uni-kl.de

© Springer Nature Switzerland AG 2020
V. Viana et al. (eds.), *Thinking, Drawing, Modelling*,
Springer Proceedings in Mathematics & Statistics 326,
https://doi.org/10.1007/978-3-030-46804-0_6

Sebastiano Serlio. For theatre stages, the question arose on how to spatialize the image and develop scenographic perspectives (Fig. 1). Serlio formulated the aim for a theatre stage, to develop "… a scene, where we see in a small room, done by the art of perspective, marvelous palaces, temples very broad, from far spacious places, adorned with various buildings, straight and long streets crossed by other routes…." [2, p. 48, translated by C.L.].

A perspectival transformed stage design with seven different view directions by Vicenzo Scamozzi in the *Teatro Olimpico* in Vicenza had been accomplished in 1585 and can be visited still today. To understand the transformed stages and explore the relation with the viewer, the stage concepts of Sebastiano Serlio und Vicenzo Scamozzi in Sabbioneta had been rebuilt in 3D models by one of our students (Fig. 2).

Other important contributions in theatre stage design as scenographic perspectives had been done by Guidobaldo del Monte [3] and Andrea Pozzo [4].

Joseph Furttenbach applied those theatre stage designs around 1640 in Ulm, Germany, with some similarities to Serlio (Fig. 3), but refined by him with prisms for changeable stages [5–7].

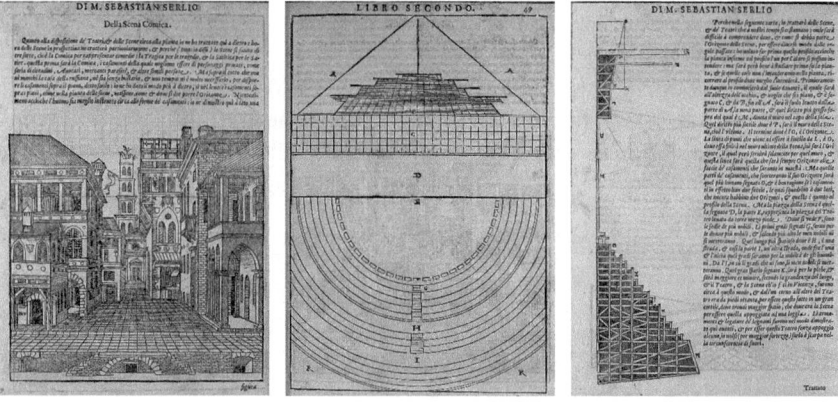

Fig. 1 Stage design by Sebastiano Serlio [2]

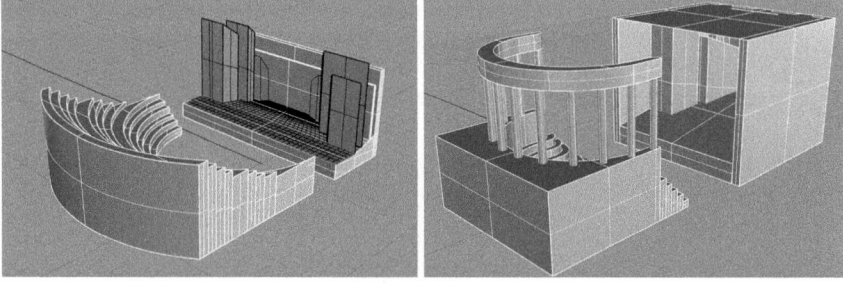

Fig. 2 3D models of the stage concept by Sebastiano Serlio and Scamozzi in Sabbioneta by Viyaleta Zhurava, TU Kaiserslautern

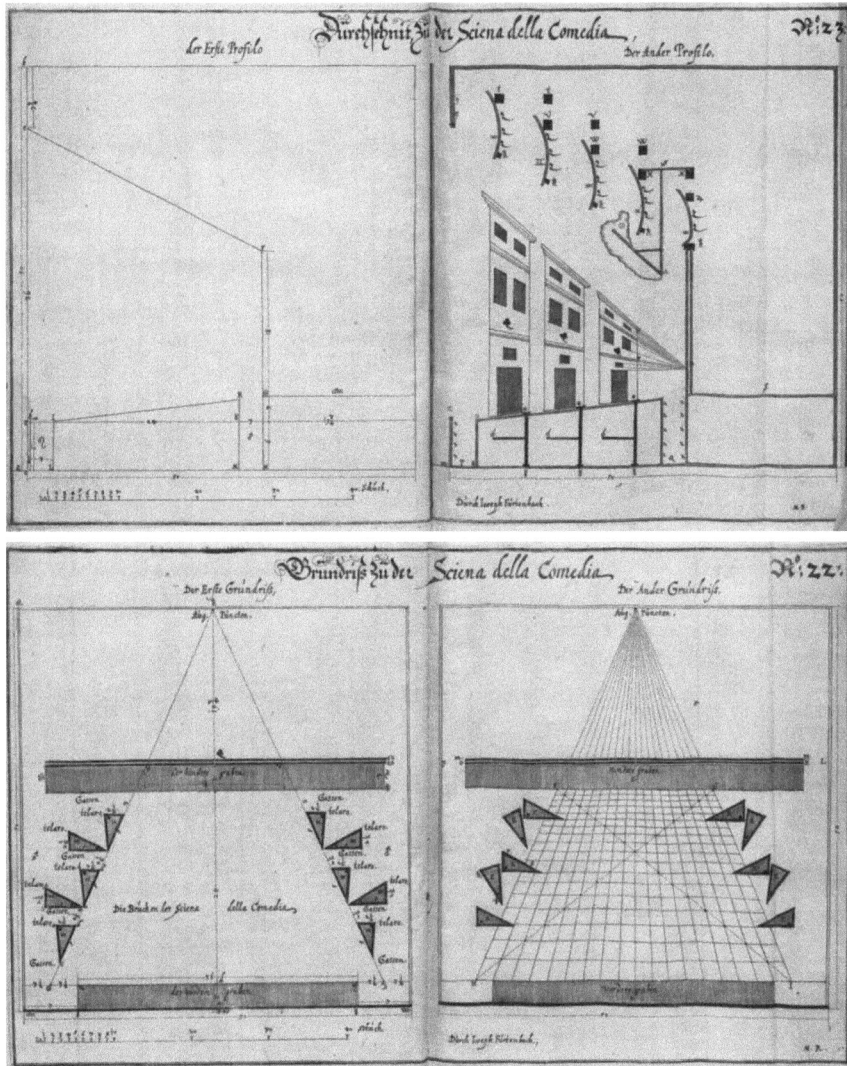

Fig. 3 Example of Furttenbach's Stage Design [6, p. 219, 222]

These had been practical concepts for stage design without the mathematical background of a transformation of spatial objects. There, we can study the idea of spatial design related to the vision of the audience. The well-known built architectural examples like Borromini's gallery for the *Spada Palace* in Rome (1635), in collaboration with Bitonti, had been thoroughly analysed by Giuseppe Fallacara and Nicola Parisi [8] and discussed with regard to relief perspective by Giuseppe Amoruso [9]. We will follow these early ideas of spatial architectural designs, related to the vision of their systematic geometric concepts as relief perspectives related to projective geometry, as well as some experiments on possible applications in architectural design.

2 Mathematical Concepts of Vanishing Points and Transformation

There are two important further developed concepts in Mathematics for the understanding of perspective as a transformation, that ended up by enabling finally spatial transformations applied on spatial architectural designs.

The first important step had been the mathematical concept of vanishing points. Guidobaldo del Monte [3] was the first to introduce "*punctum concursus*", then Brook Taylor [10] introduced vanishing points and lines in a first clearly mathematical understanding. The final work out of projective geometry with the introduction of points and lines at infinity is found in the work of Karl Georg Christian von Staudt's *Geometrie der Lage* [11], 1847, in which finally were laid the basis that all points and lines find their counterparts in space and projective transformed space, as well as in the reversed direction: "Two straight lines, laying in one plane, have either one common point or a common direction. Two different planes have either a common straight line or a common position" [11, p. 23, translated by C.L.].

The second step was the mathematical concept of perspective as a transformation of figures. For the enhancement of geometry itself, the properties of figures and their projective transformed versions had been investigated. The *Perspectograph* of Johann Heinrich Lambert from 1752 [1, p. 230] visualized already the transformational concept of plane figures. It had been further worked out in Mathematics as the projective relationship between geometric figures in plane and then also in space. Jean-Victor Poncelet and Jules de la Gournerie founded this understanding with their works. Jean-Victor Poncelet wrote *Traité des propriétés projectives des figures*, 1822. In the *Supplément sur les propriétés projectives des figures dans l'espace*, he dealt with the application of his idea of the projective characteristics on figures in space. He wrote "The homological figures must be relief projections from one another." [12, p. 358, translated by C.L.]. The idea of transformation of the spatial figures was clearly described, but there were no drawings to support practical applications. Jules de la Gournerie noted in his book *Traité de perspective linéaire: contenant les tracés pour les tableaux, plans et courbes, les bas-reliefs et les décorations théâtrales, avec une théorie des effets de perspective*, 1859, book 9, about the theory of relief perspectives, that: "Homologic transformation is the solution of the problem" [13, translated by C.L.].

The comprehension of perspective as a transformation in the frame of projective geometry laid the fundament of a systematically geometric concept of relief perspective as a transformation of spatial objects.

3   From Scenography to Relief Perspective

Only with the full mathematical understanding of vanishing points, lines and plane and of projective transformation of spatial systems, the concept of relief perspective could be developed.

Johann Adam Breysig, professor for art in Magdeburg, introduced in his book *Versuch einer Erläuterung der Reliefperspektive, zugleich für Mahler eingerichtet* [14], 1798, the word relief perspective, *"Reliefperspektive"*, that, prior to this, was more generally called as bas- or haut-relief or *"Bildner-Perspektive"*. *"Bildner"* had been another German word for "sculptor", at that time. The expression relief perspective would be more reasonable for the subject, he decided, because not all reliefs are perspectival. He claimed to have formulated the mathematical rules for a relief perspective in his book, which he did not find in other books at that time. This was probably the first book about relief perspective. The concept of vanishing points, lines and plane is the basis for formulating the rules. Although he calls the vanishing plane the main plane (the vertical plane through main point H in Fig. 4, left), the idea gets already clear that this main plane, or better called vanishing plane, determines the depth of the relief perspective, which must be set out at the beginning, for the choice of the parameters. It gets clear in his drawing that the half infinite space behind the front plane, which Breysig calls the image plane, in order to make the comparison with a perspective drawing, is transformed into a space layer between the front plane and the vanishing plane, in Fig. 4 (left) illustrated by the cuboid *IKLMNOPQ*. He applied the rules not only to spaces but also to spatial objects, as in the example of the pyramid in Fig. 4 (right), which shows a stronger transformational understanding. He showed, in another table, the relief perspective transformation of a cylinder.

The starting point for relief perspective has been accomplished in art, and then mathematically worked out on the foundation on vanishing elements and projective transformation as described. After these mathematical inventions, several authors wrote important works on relief perspective with its geometric background, as for

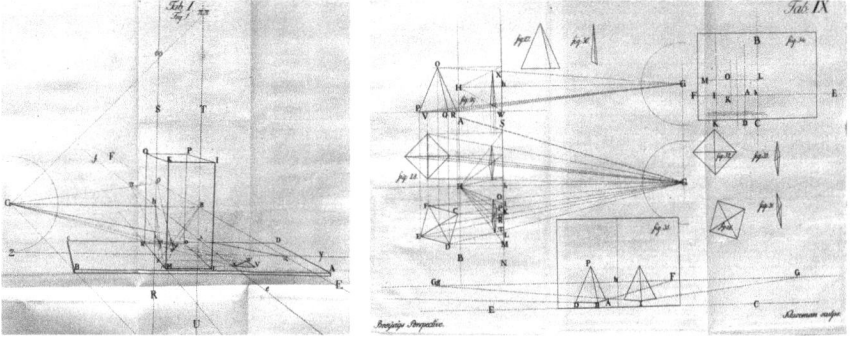

Fig. 4 Perspective drawing to explain the difference between a perspective and a relief perspective (left); construction of the relief perspective of a pyramid (right) by Breisig [14, Tab. I and IX]

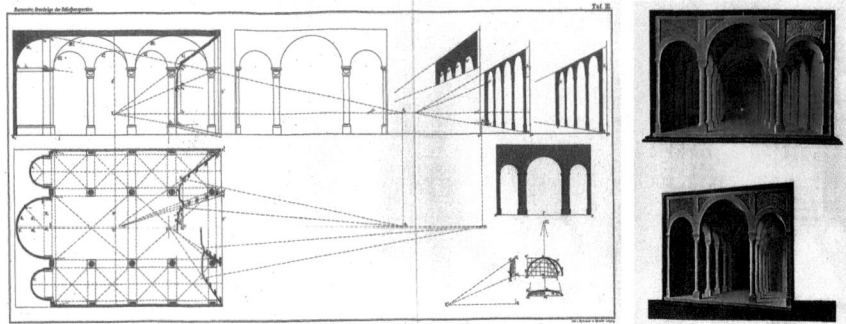

Fig. 5 Construction drawing and model of the relief perspective of a roman basilica by Burmester [17, Tab. III and IV]

example, the book of Noël-Germinal Poudra *Traité de perspective-relief* [15], 1860. He drew many examples of relief perspectives and explained a method where he started with an axonometric drawing of an object, to achieve its relief perspective.

Rudolf Staudigl wrote in 1868, *Grundzüge der Reliefperspektive* [16]—a comprehensive textbook with explanations of the theoretical background in mainly axonometric drawings. He formulated clearly the concept of collinear or homological transformations as described by Victor Poncelet and Jules de la Gournerie. He stated that "the relief perspective and its original form two perspective spatial systems, which are collinear related spatial structures" [16, p. II, translated by C.L.].

It is the merit of Ludwig Burmester, to finally combine the mathematical theory of relief perspectives with the realization of those models. He wrote in the introduction of the book *Grundzüge der Reliefperspective nebst Anwendung zur Herstellung reliefperspectivischer Modelle*, 1883: "For there is really a need to bridge the wide gap between theory and practical design" [17, p. III, translated by C.L.]. In some contributions for the *"Allgemeine Bauzeitung"*, Burmester explained relief perspective under the title *Grundlehren der Theaterperspektive, 1884* [18] with drawings in three plates for a wider audience. Figure 5 shows a constructional drawing and model of the relief perspective of a roman basilica, which was remodelled by student Giacomo N. Ghobert, Iuav Venezia, shown in Fig. 6. The original models of Burmester had been probably destroyed during the Second World War in Dresden, but rebuilt with the help of 3D printing by Daniel Lordick in 2004 [19]. After the hype on relief perspective in the nineteenth century, combined with the development of projective geometry, relief perspective had been more or less forgotten. Hermann Schaal, my geometry professor at Stuttgart's University, reconsidered the topic in 1981 in the journal *Der Mathematikunterricht* [20], applied it to architectural objects and showed a combination of perspective drawing and relief perspective model.

Relief perspective in the frame of projective geometry transforms a half infinite space behind a defined front plane to a space layer between the front plane and the set vanishing plane. If the space layer has the depth zero, we receive the normal perspective on an image plane.

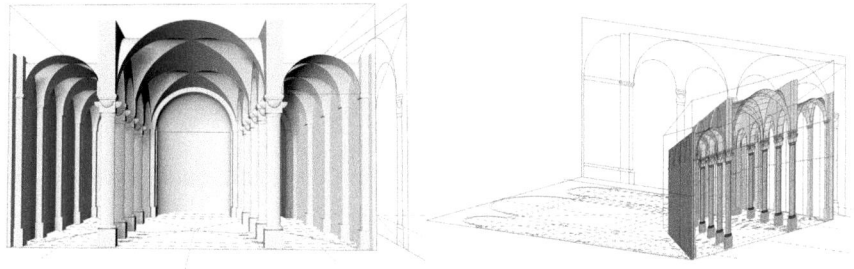

Fig. 6 3D model of the relief perspective of a roman basilica according to Burmester by Giacomo N. Ghobert, Iuav Venezia

Staudigl [16] had been aware of the general role of relief perspective in the system of projection. He expressed that the relief perspective can be understood as the most general method of projection, from which the orthogonal, the oblique and the perspective projection on an image plane arise as special cases.

4   Perception and Architectural Design

The described concept of relief perspective as a transformation of spatial objects gives the chance for applications in architectural design processes.

The analysis of the relation between different kinds of geometries and architecture by Robin Evans [21] touches an interesting point. He related the metrical geometry to haptic space and projective geometry to the vision, the visual space, following William Ivins and Henry Poincaré. This difference is for him important and enables

> us to see why architectural composition is such a peculiar enterprise: a metric organization judged optically, it mixes one kind of geometry with the other kind of assessment. Perhaps this is reason enough for the confusion surrounding it [21, p. XXXIII].

Evans argued for a stronger inclusion of projective geometry in architecture, which refers not to thought forms, but to perceived forms. He developed the idea to design architecture out of seeing, thus relating architecture to projective geometry. This may be a way to bridge the discrepancy between spatial concept and perceptibility in the architectural design.

Ernst Mach [22, pp. 5ff.] worked about sensual perception and the relationship between the perceived space and geometry. He noted that the concept of geometric Euclidean space could only result from continuous movements by a person in the perceived space. So, the specific visual perceptions precede the geometric Euclidean space concept. James Gibson analysed profoundly our visual perception in *The Perception of the Visual World* [23], 1950. He made the distinction between the visual field and the visual world. For him, the visual field is Perspective, while the world is Euclidean. He formulates that the visual field is seen, whereas the visual world is only known.

We picked up this idea to reflect spatial design processes related to our visual perception and therefore to projective geometry, by experimenting with conditions and characteristics of perspective transformations, and consequently be aware of the connection between vision and object. Some experiments on this relationship had been done already in the project "View—Space—Image" at TU Kaiserslautern, presented in *Points of View and their Interrelations with Space and Image* [24]. Studying the relationship between image and spatial design gives the opportunity to relate design with vision. An example of a created relief perspective of Stirling's Clore Gallery, made by a student group in our so-called *All School Charrette* 2013 - Stirling Hoch3 [25], with the aim to produce the same image of the model as the perspective drawing, made by Stirling [26] in his design process, is presented in Fig. 7 together with a perspective drawing by a student group.

Another example of a student group in our "*All School Charrette*" in October 2016, where the students analysed the Yale Centre of British Art in New Haven of Louis I. Kahn, is shown in Fig. 8. The task was to build a model, which can produce an image representing a specific impression of the building. While most students built a normal model of one of the rooms, one student group created a relief perspective out of a photograph of the upper galleries in the Yale Centre and produced then a photograph of the relief model which corresponds to the impression of the starting photograph.

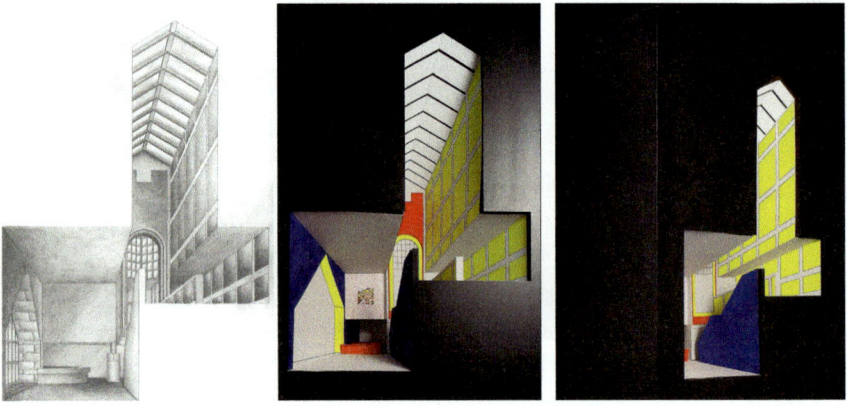

Fig. 7 Perspective drawing and relief perspective model of a student project (photographs: Bernhard Friese)

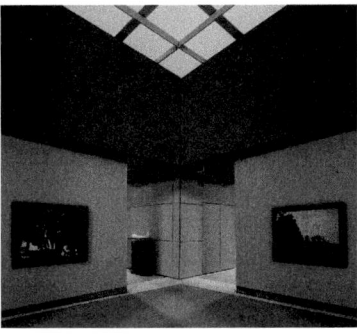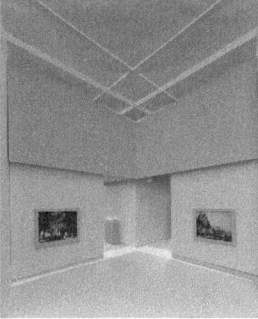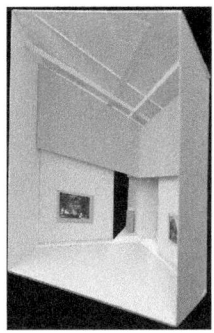

Fig. 8 Model photograph and relief perspective model of the upper galleries in the Yale Centre by
L. I. Kahn, student project, A. Rehbein et al.

5   Examples and Experiments of Perspective Transformational Approaches in Architectural Design

There are only a few examples of affine or perspective transformations in the professional practice of architectural design. The examples of Caramuel de Lobkowitz
in his *"Architectura obliqua"* [27–29], 1678, show an introduction of an oblique
transformation with the aim to adapt a building or a detail to the natural structure or location, for example, to balustrades of staircases. Claude Parent's and Paul
Virilio's [30], 1963–69, declared many years later, without referring to Lobkowitz
that *"Architecture oblique"* was important to dynamize the space and generate
activity. They stated in their *"Architecture principe"*: "In effect, the static vertical
and horizontal no longer correspond to the dynamics of human life. In future, architecture must be built on the oblique, so as to accord with the new plane of human
consciousness." [30, p. 65].

Peter Eisenman's *House X*, 1975, and *El Even Odd*, 1980, [31] can be interpreted as
based on affine transformations. The interrelations between concept, representation
and built architecture had been reflected especially in reference to Peter Eisenman's
works by Bruno Reichlin [32].

In the early work of Zaha Hadid, especially Vitra Fire Station, Weil am Rhein,
1991–93, we can explore the role of perspective in the design process, although a
perspective transformation is not strongly geometrically applied. Zaha Hadid said
in the conversation of Luis Rojo de Castro with her about the concept of Vitra Fire
Station:

> The building is made of a series of walls and planes - horizontal and vertical pieces - that
> eventually begin to shrink and expand, producing different volumes. We are interested in
> what it is like to have a projected space, a space based on projection. What would it be like
> to be in a space of resembling frozen movement? A space that seems to be in movement but
> that is also a projection, a frozen instant [33, p. 8].

The architecture of Preston Schott Cohen [34] shows a more explicit design approach on the basis of projective geometry. He uses perspective projection as a tool for the transformation of forms. In his architectural designing process, he uses perspective drawings of objects and assumes them, for example, as elevations of the object. An initial perspective of a six-sided object is assumed to be an orthographic projection, an elevation, from which the other views, a section and a plan, are later derived to produce a third dimension and again a perspective of this object. These procedures are repeated then several times. He wrote: "Thus, it is possible for a perspective or a shadow drawn on a plane surface to be inflated into a full-bodied volume as thought it had been cut from stone" [34, p. 98]. Rafael Moneo wrote in his foreword to Cohen's book: "His work as an architect takes a particularly novel turn toward the instrumentalization of projective geometry as a means by which to revel anew in the discipline of architecture". Cohen's complex design procedures are not easy to follow. They hide in their complexity the relationship between perspective and architectural design. It remains finally unclear how the resulted architecture refers to projective geometry and not only the design process. This unique approach is worth being studied more in detail, as in the attempt of Tepacevic and Stojakovic [35].

Experiments and design procedures with perspective transformational approaches, including the relationship of proportion and perspective, were the subject of our research seminar "TransForm" with students at TU Kaiserslautern, Germany, and Universitá Iuav di Venezia, Italy, in order to reflect architecture in regard to visual perception.

A project by Giacomo Magnoni, Iuav Venezia, 2017, is shown in Figs. 9 and 10. He applied two perspective transformed shapes to the building *Il Quadrilatero* in Trieste, Italy, by Rozzol Melara, Carlo Celli, 1969–82. In order to give the building a new face, capable of providing the impression of its planimetric shape in a dynamic

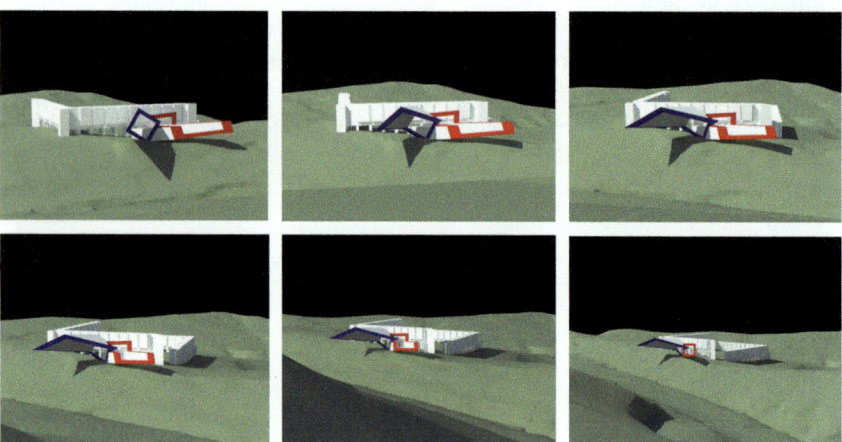

Fig. 9 Two perspective transformed shapes applied to Il Quadrilatero, Trieste—Rozzol Melara, Carlo Celli, 1969–82, project by Giacomo Magnoni, Iuav Venezia, renderings of the 3D model with views along the street

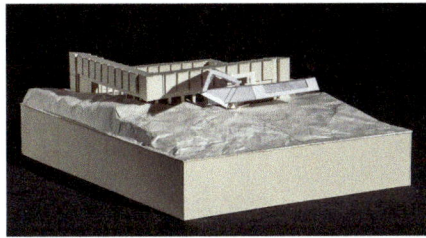 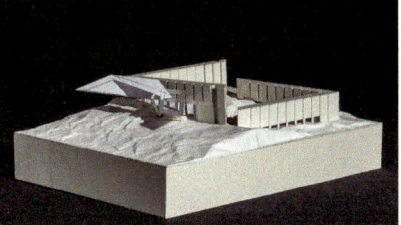

Fig. 10 Project by Giacomo Magnoni, Iuav Venezia, photographs of the physical models with the two different viewpoints to see squares

way, two projections are drawn from two points on the nearby road, and two shapes are formed, following the principles of anamorphosis. In this way, the student added an experiment on the perception of the huge building, which appears not to be adapted to the topographic situation and its perception.

Further integrations of perspectival transformations in architectural design processes are intended for future research.

6 Conclusions

The study of the history of relief perspective shows a fruitful interaction between art, architecture and mathematics. The way from theatre stage design to the foundations of projective geometry enabled us to understand relief perspective as a transformation. The works of Breysig, Poncelet, Staudigl and Burmester had been discussed in their role and classification of this development. Geometry nowadays is essentially based on the notion of transformation and symmetries as well are understood as transformational concepts. Today, the digital transformation tools give powerful possibilities to apply the concept in many ways. The perspective transformation in its spatial understanding offers the chance for integrating the perception of the viewer in architectural design processes. The tendency to get more and more complex, hides and encrypts the perception of these applied transformations. It would be worth to work further on the perceptibility of the perspective transformation in architectural designs.

References

1. Leopold, C.: Perspective concepts. Exploring seeing and representation of space. J. Geom. Graph. **18**(2), 225–238 (2014)
2. Serlio, S.: Il Secondo Libro di prospettiva. In: Tutte l'Opere d'Architettura et Prospettiva. presso Francesco de' Franceschi, Venetia (1545)
3. del Monte, G.: Perspectivae libri sex. Girolamo Concordia, Pesaro (1600)

4. Pozzo, A.: Perspectiva Pictorum Et Architectorum. Rules and Examples of Perspective. Proper for Painters and Architects, etc. London 1707, Reprint, New York 1971 (1707)
5. Furttenbach, J.: Mannhaffter Kunst-Spiegel. Johann Schultes, Augsburg (1663)
6. Furttenbach, J.: Architectura Recreationis. Johann Schultes, Augsburg (1640)
7. Reinking, W.: Die sechs Theaterprojekte des Architekten Joseph Furttenbach. 1591–1667. Tende, Frankfurt A.M. (1984)
8. Fallacara, G., Parisi, N.: Querelle di paternità. La Galleria Spada tra il Borromini e il Bitonti. Studi Bitontini **77**, 37–61 (2004)
9. Amoruso, G.: The relief-perspectives of bitonti and borromini: design and representation of the illusory space. In: Amoruso, G. (ed.) Handbook of Research on Visual Computing and Emerging Geometrical Design Tools. Hershey, IGI Global (2016)
10. Taylor, B.: Linear Perspective: Or a New Method of Representing Justly All Manner of Objects as They Appear to the Eye in All Situations. R. Knaplock, London (1715)
11. von Staudt, K.G.C.: Geometrie der Lage. Friedrich Korn, Nürnberg (1847)
12. Poncelet, J.-V.: Traité des propriétés projectives des figures. Paris. 2nd edn, in 2 volumes, 1862, 1865 (1822)
13. de la Gournerie, J.: Traité de perspective linéaire: contenant les tracés pour les tableaux, plans et courbes, les bas-reliefs et les décorations théâtrales, avec une théorie des effets de perspective. Dalmont et Dunod. Mallet-Bachelier, Paris (1859)
14. Breysig, J.A.: Versuch einer Erläuterung der Reliefperspektive zugleich für Mahler eingerichtet. Georg Christian Keil, Magdeburg (1798)
15. Poudra, N.-G.: Traité de perspective-relief. J. Corréard, Paris (1860)
16. Staudigl, R.: Grundzüge der Reliefperspektive. Seidel & Sohn, Wien (1868)
17. Burmester, L.: Grundzüge der Reliefperspective nebst Anwendung zur Herstellung reliefperspectivischer Modelle. B. G. Teubner, Leipzig (1883)
18. Burmester, L.: Grundlehren der Theaterperspektive, Allgemeine Bauzeitung 49, Wien. 39–40, 44–49, 53–57, Tables 25–27 (1884)
19. Lordick, D.: Reliefperspektivische Modelle aus dem 3D-Drucker. *IBDG* 1/2005, Innsbruck. 33–42 (2005)
20. Schaal, H.: Reliefperspektive. Der Mathematikunterricht, 27, 3/1981, Stuttgart. 69–90 (1981)
21. Evans, R.: The Projective Cast. Architecture and Its Three Geometries. The MIT Press Cambridge, Cambridge, Massachusetts (1995)
22. Mach, E.: Space and Geometry. In the Light of Physiological, Psychological and Physical Inquiry. The Open Court Publishing Company, Chicago (1906)
23. Gibson, J.J.: The Perception of the Visual World. Houghton Mifflin, Boston (1950)
24. Leopold, C.: Points of View and their Interrelations with Space and Image. In: Vera Viana et al. (eds.) Geometrias & Graphica 2015. Proceedings, Universidade Lusíada de Lisboa. vol. 1, pp. 365–373. Aproged, Porto (2016)
25. Leopold, C.: Reliefperspektive. In: Loos Asplund Stirling Diener Hoch3, 1–4. Hoch3 Stirling. Park Books, Zürich (2014)
26. Stirling, J., Wilford, M., et al.: Buildings and Projects 1975–1992, vol. 104. Verlag Gerd Hatje, Stuttgart (1994)
27. de Lobkowitz, J.C.: Architectura civil recta y obliqua. Vigevano (1678)
28. Navarro Morales, M.E.: Architectura Civil Recta Y Obliqua: A Critical Reading. McGill University, Montreal (2012)
29. Iurilli, S.: Trasformazioni geometriche e figure dell'architettura. L'Architectura Obliqua di Juan Caramuel de Lobkowitz. Firenze University Press, Firenze (2015)
30. Johnston, P., (ed.): The function of the oblique. The architecture of Claude Parent and Paul Virilio 1963–1969. Architectural Association (AA documents/Architectural Association; 3), London (1996)
31. Eisenman, P.: Eisenman Inside Out. 1963–1988 Selected Writings, pp. 144–151. Yale University, New Haven London (2004)
32. Reichlin, B.: Reflections. interrelations between concept. Representation and Built Archit. Daidalos **1**, 60–73 (1981)

33. Hadid, Z.: Vitra fire station, Weil am Rhein, 1991–1993. El Croquis Zaha Hadid 1992–1995, **73**(1) (1995)

34. Cohen, P.S.: Contested Symmetries and other Predicaments in Architecture. Princeton Architectural Press, New York (2001)

35. Tepavcevic, B., Stojakovic, V.: Representation of non-metric concepts of space in architectural design theories. Nexus Netw. J. **16**, 285–297 (2014)

Cornelie Leopold is teaching and researching in the field of architectural geometry at fatuk, Faculty of Architecture, Technische Universität Kaiserslautern, Germany, as academic director and head of the section Descriptive Geometry. She received her degree in Mathematics and Philosophy at the University of Stuttgart, Germany. She is member of the Editorial Board of the *Journal for Geometry and Graphics*, of the Scientific Committee of the Journal *Disegno*, and since 2019 corresponding Editor of *Nexus Network Journal. Architecture and Mathematics*. She was founding President of the *Deutsche Gesellschaft für Geometrie und Grafik* (DGfGG). She has participated with lectures, papers, and reviews in many international conferences, journals and books. In 2017, she was Visiting Professor at Università Iuav di Venezia. In 2018, she co-organized *RCA, Research Culture in Architecture - International Conference on Cross-Disciplinary Collaboration* at fatuk, and in 2020, the conference *NEXUS 2020: Relationships of Architecture and Mathematics*. www.architektur.uni-kl.de/geometrie; www.researchgate.net/profile/Cornelie_Leopold

Ordered Creativity: The Sense of Proportion in João Álvaro Rocha'S Architecture

Joana Maia and Vítor Murtinho

Abstract This paper aims to understand the value of proportion (in the sense of its classical tradition) as a tool for the architectural project, and the meaning that it assumes today in the course of project's methodology within the context of Portuguese architecture. Presenting João Álvaro Rocha as a case study, two of his most representative buildings will be analysed in respect to his methodological practice. The result will allow us to understand, not only the content, but also the importance that this kind of tools has, even today, in the project composition, as well as its framework on the contemporary context.

Keywords Proportion · Geometry · Number · Coherence · Order

J. Maia (✉)
Departamento de Arquitectura Da, Universidade de Coimbra, Coimbra, Portugal
e-mail: joana.maia@uc.pt

V. Murtinho
Departamento de Arquitectura and Centro de Estudos Sociais Da, Universidade de Coimbra, Coimbra, Portugal
e-mail: vmurtinho@uc.pt

© Springer Nature Switzerland AG 2020
V. Viana et al. (eds.), *Thinking, Drawing, Modelling*,
Springer Proceedings in Mathematics & Statistics 326,
https://doi.org/10.1007/978-3-030-46804-0_7

1 Introduction

Francisco Mangado defines João Álvaro Rocha (JAR) [1959–2014[1]] as a coherent architect [1].[2] This notion of *coherence* derives largely from a sense of *precision*, not only formal and conceptual[3] but mostly methodological, that characterizes the work of this architect. The rigour appears as a *modus operandi* that is, necessarily, always present,[4] consolidating a methodical attitude objectified in the implementation of a logical composition, where the importance of the inter-relationship between parts seeks, as an end, a unified and balanced solution. The methodology adopted by the architect calls upon the articulation of several classic proportional systems, from the most intuitive to the most elaborate form, used freely because they are consequent of the inherent specificities of each project. The geometric and arithmetic proportions are constant, from the creative process to the constructive implementation of the building, a systematization that will structure the justification of the drawing decisions. The importance of the method to the quality of the project/building demands a careful attention of a theme that nowadays seems to have been forgotten in theory, not in everyday practice.

2 Methodology

Graduated from the School of Fine Arts of Porto in 1986, the architect JAR mingles very closely with prominent personalities of that period, seeing himself strongly influenced by the so-called Porto School. Sensitive to the principle "the idea is on the site",[5] he also adopts drawing as a primordial design tool, at the same time that he pursues rigour in the constructive process. This fact probably derives, not only from Álvaro Siza's construction classes, but also from the practice of *atelier* with the architect Jorge Gigante, seen at the time as "the Neufert of architects".[6] In this sense, it

---

[1] At the time of development of this research, the architect JAR was at the height of his career. After its conclusion, in 13 September 2014, illness interrupted prematurely a career already recognized as solid and still promising. The initial interview and the timely collection of archive material, as well as subsequent meetings with his collaborators and the opportunity to show the results to his wife (also an architect, with a recurring critical role in the studio), allowed not only the natural conclusion of this research but also its consequent validation in general.

[2] It is not only Francisco Mangado [1, p. 4] who makes this reference. This characterization is shared by authors such as Ricardo Merí de la Maza [2], perceiving an "eagerness of justification" as Carlos Nuno Lacerda Lopes affirms [3, p. 14]. Nevertheless, Ricardo Merí de la Maza [4] goes a step further, by emphasizing the complexity and contradiction hiding behind this same formal coherence that, however, results in a synthetic and integrator tension visible in some of his works.

[3] João Álvaro Rocha [5, p. 13] highlights the importance of the *coherence* factor as a form of relationship between content and form.

[4] See the discourse of João Álvaro Rocha himself [6, p. 159].

[5] A *dictum* that characterizes the School of Porto, by the way of Álvaro Siza Vieira.

[6] Idea expressed by João Álvaro Rocha himself [7].

is not surprising that the critic's review emphasizes a posture of *coherence, precision* and *rigour* in JAR, a quest for the essential value easily tunable to the proportional universe. Although prematurely departed, this architect from the city of Maia ends up leaving, in the example of his work and methodology, a special affection for this concept without which, moreover, the work cannot be fully explained and understood. "Proportion is in everything [...], that coherence is mainly due to proportion, under the penalty of, suddenly, [the work] earning a foreign body",[7] says JAR [7], for whom the subject being, obviously, not everything, still is inescapable in the project practice and in the way of understanding architecture. His discourse is unequivocal, obsessive for being passionate, coherent in the justification of the use of a fundamental tool to the encounter between creativity and order: the pillars that support the act of composition. The posture, silent (but even so, not comfortable[8]), seeks the balance between emotional and rational in the exact *misura* that attributes commensurability to the idea, without loss of the aesthetic/artistic thinking.

Philosophically, JAR associates *proportion* to the notion of balance, in the search for a sense of perfection that belongs to the domain of the absolute, not by its meta-physical character, but by that of the unattainable: it is in the constant improvement of this infinite sequence (in its most diverse perspectives) that the great goal of art, after all, resides. In practice, he argues that the Greeks where those who came closest to this idea of perfection, standing, to this day, as a major reference. Yet, he simultaneously opens doors to its development in the present day, where the concept is expressed in an elastic constant that widens the possibilities to asymmetry, movement and even opposites conciliation. Although critical of formal exploration per se, as an adept of a logic of continuity, he does not deny the creative value that is, in fact, an essential motivation for the operation of the tool. Sterile in itself, the instrument demands a creativity not uprooted, but above all focused on the establishment of new relations: it is the prospect of proportion not as a limitation but as a guideline, in which an idea of *formulation* is superimposed to that of *formula*. So, the desired balance or the search for a sense of unity in JAR intends not to seek an archetypal unity or a priori solutions (often seen as simplistic), but it flows in interrelationships throughout the project process by constructing a story.[9] A narrative does not end in the compositional process, but extends to the several procedural phases: program, conception, detail and materialization. Several systems and strategies intersect witnessing that it is in the control of diversity, in parallel with the project theory and the confrontation with

---

[7]Free translation from: "A proporção está em tudo [...], essa coerência passa muito pela proporção sob pena de, de repente, [a obra] ganhar um corpo estranho".

[8]Idea advanced by Antonio Ravalli [8, p. 6]. Ricardo Merí de la Maza shares this thought by calling the process of João Álvaro Rocha a "quiet revolution" [2].

[9]"I am increasingly interested in focusing on the ideas, on the plane of intentions, and verify whether they actually materialize, if they take shape in a coherent way and less and less interest me [...] the formal and/or spatial *a prioris*, which often prevent us from perceiving what is important". Free translation from: "[...] cada vez mais me interessa fixar-me nas ideias, no plano das intenções, e verificar se elas se concretizam efetivamente, se tomam corpo de um modo coerente, e cada vez menos me interessam [...] os a priori formais e/ou espaciais, os quais nos impedem muitas vezes de perceber o que é importante" [6, p. 160].

the urban tissue, that the contemporary project must be worked once implemented in a postmodern context. The act of commensurability of the idea is already present in the initial thought and gains space throughout the composition process, participating actively in the making of an exercise of ordered creativity. Being architecture an art linked to social service, order becomes peremptory as a support of a proportional response to the purpose that is required of it. In this sense, the dimensions of the theme rise to several fields of the project exercise, measuring the adequacy of decisions. And since architecture only becomes fact when it is built, the same level of attention that is applied to the method of conceptual and physical composition should be given to the constructive process: this is the difference between *design* and *drawing*.[10] For JAR, the constructive details and all the logic associated with them must be inserted immediately in the initial phase of ideological approach, safeguarding the operative possibility of the project act: it does not matter the genius of an idea if it does not have the capacity to be put into practice. Only in this way, it will be possible to reach the intended assumptions, structuring a logical and intelligible discourse for communication between the various agents involved.

Notwithstanding this theory, there is nothing better than analysing the work itself. To discover methodological approaches and, above all, to understand how they move from thought to the core of practice: its adjustments, its subtleties. In this perspective, we propose the analysis of two fundamental works of JAR that, starting from distinct premises, help us understand the *act* in itself.

3   Case Study #1—Paçô House, Viana Do Castelo (1994–1997)

The first example, being a construction made from scratch, presents a development with an intimate relationship with the surrounding territory. Laid out in the manner of a promontory, a large longitudinal window embraces a diversity of volumes geometrically and materially interrelated, in an attempt to establish a relation to the preponderant axis of the place: the horizon line defined by the sea, parallel to the direction of the contour lines of the terrain. This initial regulatory line paves the way to, through relations of parallelism and perpendicularity, implement three fundamental axes that outline the basic structure of the complex, developing the three main moments of the composition: the volume of the house, the body of the pool (both parallel to each other), and the path that connects them and establishes a relationship with the exit of the intervention area (Fig. 1).

Within a lot of irregular geometry, some notable points of the perimeter wall give the commanding lines for a justified development of the composition, with the external irregularity contributing to the drawing of the internal order. The point $P1$,

---

[10]João Álvaro Rocha [7] makes a distinction between *design* and *drawing*, highlighting the universal, aesthetic and free strand of the first, in counterpoint to the restrictive nature of the second, centred on responding to previous constraints that are characteristic of the profession.

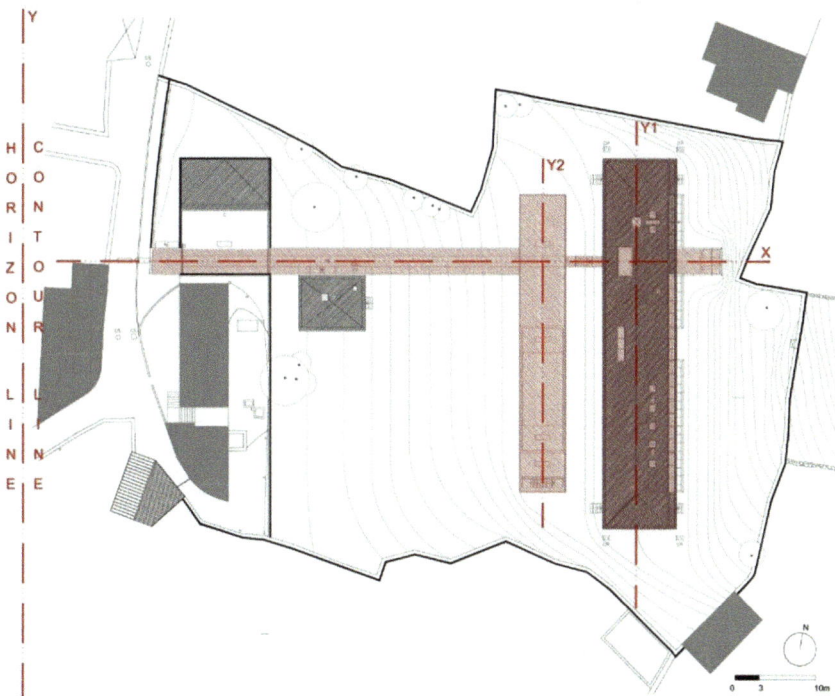

Fig. 1 Masterplan. General guidelines

vertex of the boundary wall, is the reference point for the definition of the dominant axis (straight line *a*), which subsequently limits the housing volume. Point *P*2,
denoting the moment of entrance in the lot, allows the drawing of the perpendicular
axis that materializes the access route to the house (straight line *b*). From these axes,
parallel lines are multiplied in order to define the necessary alignments (Fig. 2).

The front limit of the housing volume relates to vertex *P*1′ of the pre-existing
neighbouring structure, as its north and south alignments establish relations, respectively, with the annex and the pre-existing construction in the western side. From point
*P*3, a new line, parallel to the dominant axis, accompanies the western boundary of
the pre-existing construction to define, in the convergence with line *b*′, point *P*4,
vertex of the first annex. The length definition of the internal volume, coinciding
with the depth of the pool platform, is determined by the straight lines *b*1 and *b*1′
(parallels to axis *b*) which depart, respectively, from points *P*5 and *P*3. Now, if point
*P*3 is a point with a previous definition, point *P*5 will result from the intersection of
the straight lines *c*′ and *c*1, perpendicular to each other and with an angular dimension
of 45° with respect to the direction of the initial regulating lines, axes with origin
in points *P*2 and *P*4. The game begins in the parallelism of the straight line *c*′ with
respect to the line *c*, an axis that, starting from point *P*1, crosses the lot in a diagonal
form. Passing through fundamental points such as the south-west vertex of the pool

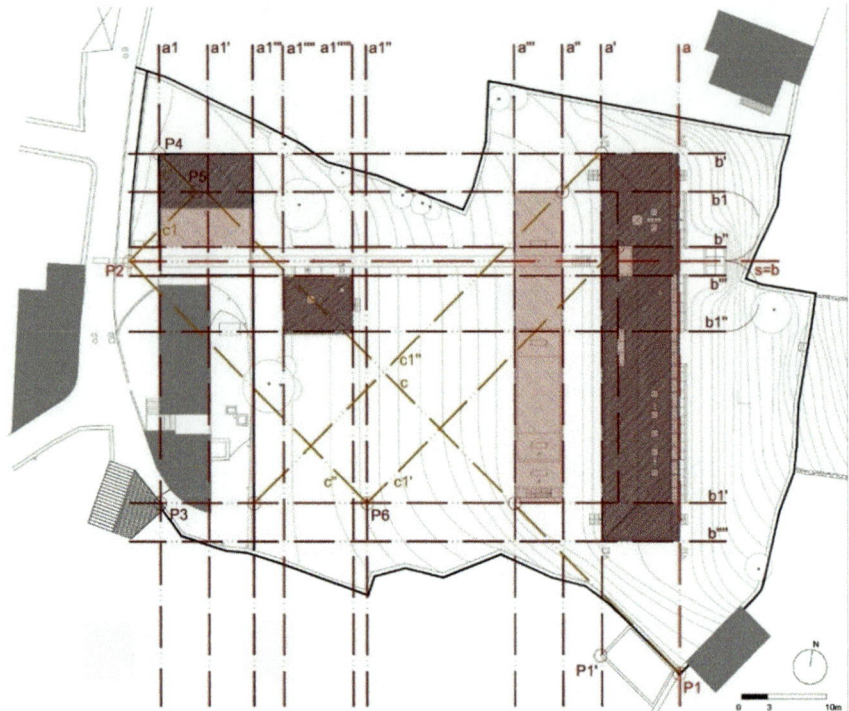

Fig. 2 Masterplan. Regulating lines

or the central point of the second annex, line *c* meets line *b*1 at the point of intersection with line *a*1′, a coincident axis with the internal alignment of the pre-existing
constructions. The retraction of the housing volume in relation to the west alignment
of the external body is established by the convergence of line *b*″ with line *c*1′ . Line
*c*1′ arises by symmetry to line *c*″ (originating from point *P*2), by way of the axis *a*1″
which is established with origin in another vertex of the irregular perimeter of the
terrain, travelling parallel to the dominant longitudinal axis up to point *P*6 (in the
intersection with line *b*1′). At the same time, the straight lines *a*″ and *a*‴ parallel to
line *a* and with relation to lines *c* and *c*1″ (perpendicular to line *c*) outline the longitudinal alignments of the pool. The alignment of the beginning of the pool itself (line
*b*1″) is marked by symmetry to the north alignment of the platform (line *b*1), with
respect to the axis of reflection symmetry *s*, coincident with the initial alignment *b*.
Both annexes of the house, located west of the plot of land, also end up being defined
in the course of the geometric handling, adopting alignments that are parallel to the
fundamental axes, thus contributing to set the transversal dimension of the access
route to the house. A simple perpendicular alignment to the first diagonal, line *c*1″,
confirms the relationship between different points of the composition. JAR's study
drawings (Fig. 3) reveal the attempt to synthesize this subliminal geometric structure
that pins the proposed object in the premises of the existing context. An intuitive

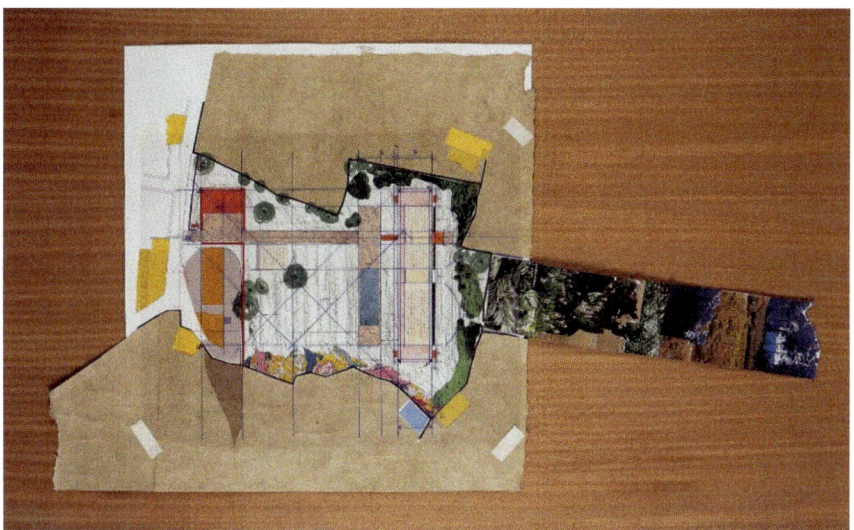

Fig. 3 Synthesis scheme for masterplan

work of symbiosis with the place defining, in the first instance, the composition, to later concentrate on the compositional development of the object itself, culminating in a third phase of detail improvement.

Moving on from the broad plan to the architectural craft, the thought of a clear metric is immediately expressed in sketch drawings, an intention confirmed by the drawings of the project (Fig. 4).

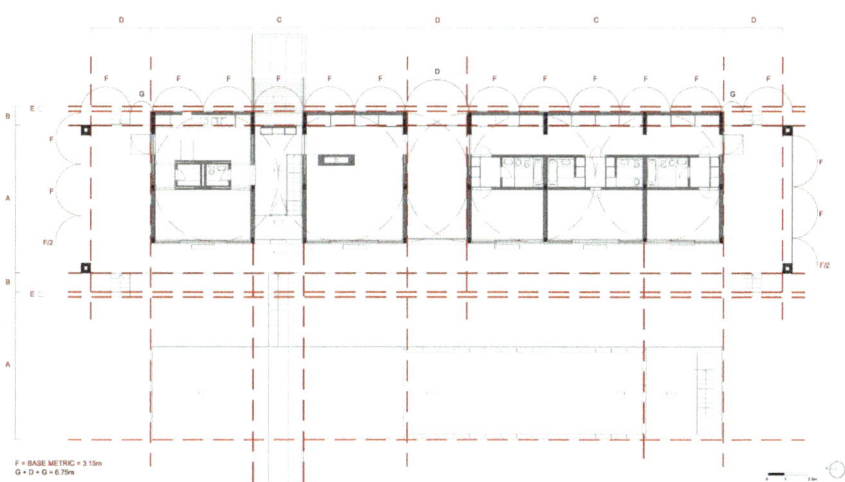

Fig. 4 Ground floor plan. Metric

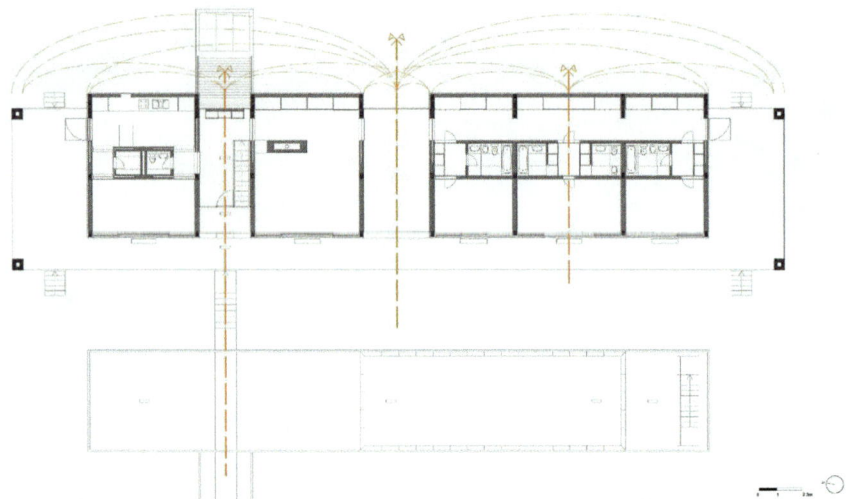

Fig. 5 Ground floor plan. Reflection symmetry system

In relation to the motion of the longitudinal axis, the metric consists of a sequence of type *BABA,* where *B* is the average depth value of the platform access stairs, and *A* the reference dimension of the side view of the house's external volume. However, the difference in elevation expressed in the topography did not allow the exact implementation of the initial metric, forcing the elimination of one step on the east side and its addition on the west side (dimension *E*): the posterior alignment of the internal volume also follows that imposed modification. Regarding the displacement of the transverse axes of the volume, the metric of key alignments assumes a *DCDCD* sequence. Despite the apparent volumetric fragmentation that the project preserves inside a unitary outer shell, it does not become disconnected from a sense of proportional unity. The symmetry also appears as a tool, reinforcing the control of the fractures that are intentionally integrated (Fig. 5).

The central axis of reflectional symmetry tears the unity of the inner body materializing a meeting room, which assumes the separation between the social area and services (to the left) from the private space (to the right). Each of the two resulting bodies has the same *C* dimension, yet they express themselves differently. If the south block maintains the volumetric integrity, the northern block is allowed to be teared similarly to the global volume.[11] Assuming a central opening (entrance) coincident with the outer access axis, the gesture establishes the division between service and living areas and is extended by the floor design that denounces the laundry space in the sub-floor. The strategy of centrality extends to the detailed drawing as, for example, in the positioning of bloc steps of access to the platform in relation to the openings. At the same time, methods of animation of the staticity are applied, as in

---

[11]Despite the integrity of the southern bloc, it is noteworthy the strategical similarity between this and the northern bloc through the use of reflectional symmetry for the construction of the sector.

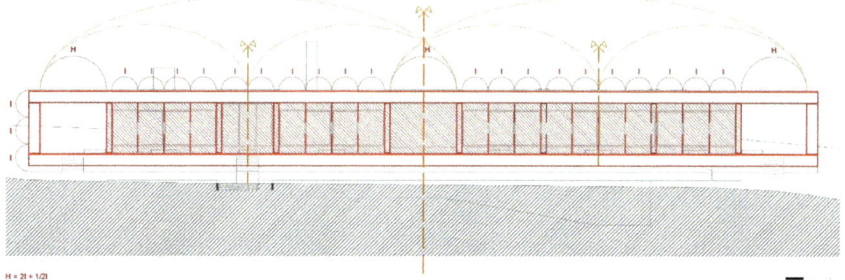

Fig. 6 West elevation. Metric; symmetries; relation between elements

the movement of the internal volumes in the east direction allowing, simultaneously, the design of the balcony space. The analysis of the final project drawings reveals a metric based on the F module, arithmetically corresponding to the value of 3.15 m, that is, an arithmetic proportion based on the number three system (Fig. 4). Three moments set themselves apart from the base metric: the central dimension D and the two sides G; however, its sum results in 6.75 m, integrable in the system. Although the longitudinal cadence is the most valued, the transverse relation does not cease to exist, being integrated in the adopted logic. Also, noticeable is the relation of (4 + 1/2): 1 in the composition of the internal body, where the volumes articulation is divided in the central axis for doubling the double square.

From the altimetric point of view, it is possible to see the rhythm printed on the metric of the openings (value I), in direct relation with the three highlighted openings (value $H =2I +1/2I$) (Fig. 6). The height of the inner space of the outer platform, more conditioned with the anthropometric relation, makes an approximation to the value of 2.70 m in order to give continuity to the base system. In the plane of the facade, the relations of symmetry are equally in evidence: in the big scale, as well as in the scale of the detail. In the analysis of the drawing, the importance given to the independence of volumes reveals itself as evident. The same thing happens in relation to the care for the dimensional constancy of the surrounding piece thickness, reinforced with the precise underline on the continuity of the base.[12] This allows a clear reading of the articulation between parts, as well as a greater identity of the external structure. A systematization orchestrated for greater "cleanness" of the drawing, developed also in the phase of the enhancement of the detail.

---

[12]The solution is based on an ancient technique: a ceramic piece of 3 cm draws a horizontal line that surrounds the whole volume and makes the separation between volume (fixed dimension) and base (lower dimension, variable). In spite of the small size of the ceramic piece, this kind of cutline ends up being a sufficient element to the necessary differentiation between parts, thus aiming for a clear reading of the volumetric proportion.

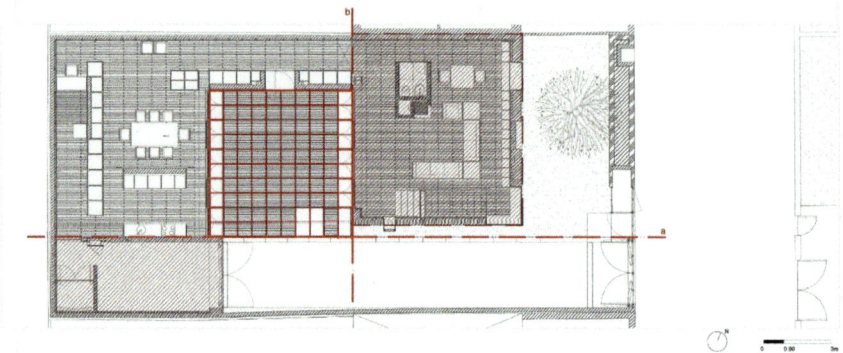

Fig. 7 Ground floor plan. Regulating lines; geometry and courtyard modulation

4   Case Study #2—Tomé Sousa House, Porto (2001–2009)

As a rehabilitation and expansion project, this second example reveals specific characteristics, in which the parameter *place* crystallizes itself mainly in the premises of the pre-existing object. With the lot initially constituted by housing and annex, the solution passed through a spatial continuity between these independent volumes, denoting a maximum exploitation of the posterior area of land. A new body of rectangular geometry, made up by full/empty duality, assumes the connection between the two existing spaces in the ground floor. Despite the introduction of this sense of continuity, the memory of the lot is preserved mainly by the incorporation of the courtyard that JAR denominates as the "new centre of gravity of the house". A new entrance space that agglutinates the whole composition, and whose material, spatial and proportional treatment provides a continuous reading (Fig. 7).

Although restructured, the spatial distribution maintains the pre-existing logic: social and service areas on the ground floor, releasing the private space to the upper floor. The two pre-existing volumes emphasize two regulating lines, that are perpendicular to each other. Line *a*, longitudinal to the lot and shaped in the drawing of the pavement, defines the garage space and the entire access range. Yet, line *b* transversal to the lot, defining the posterior limit of the pre-existing body of the house, establishes the boundary between the master volume and the new volumetric proposal. It is at the moment of intersection of these two axes that the courtyard space is born, with this interior/exterior mediating element of squared geometry outlining the remaining alignments of the house. The search for centrality and stability is fixed through this simple geometric figure endowed with comprehensibility, establishing a sense of unity. The introduction of steel cables to connect volumes[13] accentuates the interior/exterior ambiguity of this element: exterior extension of the living and dining rooms. The relationship between parts is made both dimensionally and materially, with the wood being prominent and conformed as modular gauge for the

---

[13]A sort of pergola waiting for the growth of the coming vegetation to unify the whole ensemble.

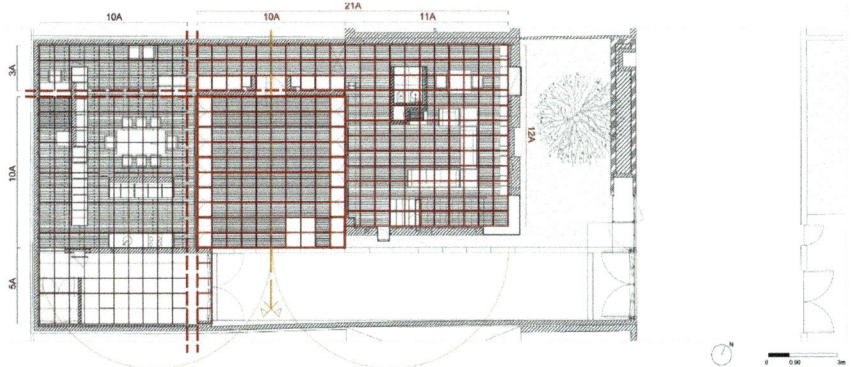

Fig. 8 Ground floor plan. Base modulation and its extension; centrality of the courtyard

dimensioning of the space, generating an image of continuity. The game develops itself on two fronts articulated between themselves: the modulation of the cabinets and portals and the alignment of the floor joints, both in relation to the pre-existing data. Dimension 0.63 m is the constructive result of a conceptual metric of 0.60 m, descending from the base square and an arithmetic ratio in the system of number three. Consolidated the physical commensurability, the composition is subject to adjustments between metric and architectural elements by inverting the logic: the stereotomy of materials gains prominence over the established grid. The moment is to subvert the rule to a greater end. The adjustment meets the pattern that factually draws the entire project space, that is, the width of 0.075 m from the wooden floor piece.[14]

Figure 8 demonstrates the articulation of the courtyard cadence with the pre-existence, and the intention to expand it to the rest of the composition. Conditioned by the boundaries of the lot, the metric loses some strength in the posterior part of the intervention, although it denotes a certain character (of alignment) in the remaining spaces. The original access to the housing is transferred to the central axis of the courtyard, a hinge that reflects a dimensional approach in a reflectional symmetry to the whole set. The ground floor is organized in a fluid way, without the use of effective spatial divisions, but incorporating simple elements that establish the organizational layout. The material and geometric continuity emphasizes the fluidity of space, with the exterior of the courtyard being integrated into the internal unit of the house.

While the ground floor is characterized by a great permeability and depth reading capable of suggesting proportional amplitude in the interior area, the upper floor, limited to pre-existence dimensions, reveals a more closed and secluded character that conforms with its private function. The composition reveals the definition of a first positioning range for the sanitary installation of the main room that, in continuity with the access ladder, releases an area of square geometry for the drawing of structural spaces (Fig. 9).

---

[14]84×0.075 m $= 6.30$ m, corresponding to 10 modules of 0.63 m, side of the courtyard square.

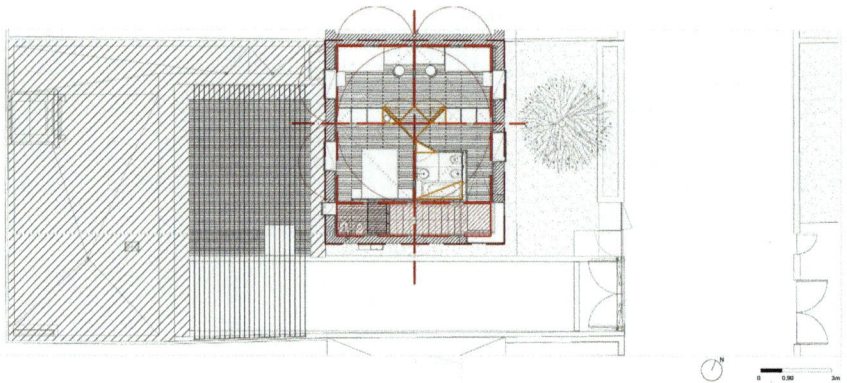

Fig. 9 Upper floor plan. Geometric base; alignments; geometric resolution of details

The strong stability of the square is highlighted when, inside this geometric definition, two axes of reflection symmetry arise as alignments. At the point of intersection of these axes, a new square rotated at 45° is suggested in the drawing of the plan defining the distributive centre. The exercise is a manipulation of the minimum dimension, where the door element defines not only the perimeter of the hall but also the dimension of passage to the different spaces. A work of scale, in which geometry contributes to the tension of the body, released at the moment of dematerialization of this geometric space. The perception of geometry arises by approximation, in the perspective of the corridor, perceived in full at the moment when the observer stands in the place of the missing arris. Triangular effects solve some of the details. As in the ground floor, we also verify the compatibilization between geometric premises and the stereotomy of materials, in the wooden floor and in the ceramic pieces of the service spaces. The care given to the alignments, the compatibility between elements and joints, or the centrality of the parts in relation to the stereotomy, are strategies used to create a dynamical unity to the composition.

In altimetry, a similar consideration for the alignments can be verified, even when the absence of materials of modular application reduces the exercise to the structural elements of the composition. The larger openings give continuity to the pre-existing ones, but they are redefined for contemporaneity in ways of treatment, both material and linguistic. When they become obsolete the memory of the opening is fixed, although deprived of its function, not only preserving the urban setting but also the balance of the facade composition itself. On the other hand, the smaller openings, defined through the geometry of the square, articulate themselves more independently although seeking relationships between each other and with the larger spans (Figs. 10 and 11). The overall result denotes a sense of rigour and preservation of the identity of each element, in harmonious relationship with the other elements: a fortunate balance between past and future.

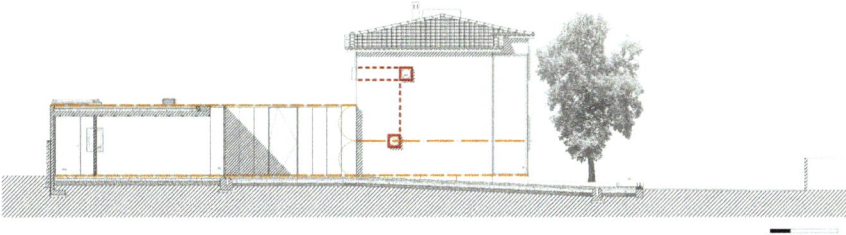

Fig. 10 South elevation. Geometries; alignments

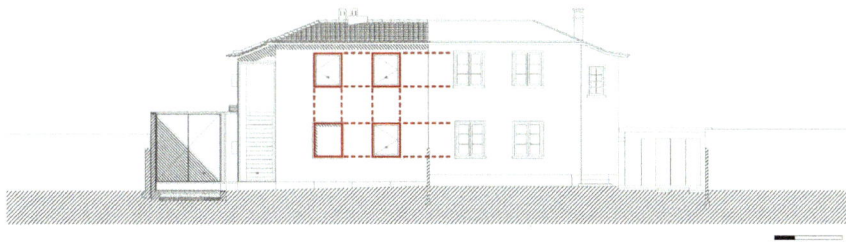

Fig. 11 East elevation. Alignments

5   Final Remarks

Both case studies reveal a proportional work developed in three distinct stages: the scale of the place, the scale of the object and the scale of the detail. The major difference in terms of strategy lies mainly in the larger scale, since the geographical references of the place (Paçô example) are replaced by the specificity of the pre-existence in the example of Porto (with all the necessary adaptation that ensues): the conditions of each project are what dictates its guidance. The use of systems is done in a free and personal way, transversal from the act of creative conception to the operative systematization of the drawing, with a permanent awareness not only of the importance of the human body, but also of a critical and permanent look fundamental to the whole procedure. Geometric and arithmetical proportion challenge themselves in a joint dance where regulating lines, reflectional symmetries, modular metrics or simple geometric figures articulate with a systemic network conceptually and factually based on the number three system, structuring a solidly ordered and justified drawing. The potential theoretical/conceptual rigidity becomes more flexible within the necessary adjustments to the real constraints, while the robustness of the assembled construction allows the safe deviation to framework in contemporaneity. The validity of the methodology adopted resides mainly in the intellectual logic that is printed in the project, configuring exercises where aesthetic value mingles with operational rigour (because clear and coherent) without rigid orthodoxy, a systematization that facilitates communication among the various agents involved.

6 Conclusions

To JAR [7] Portuguese contemporary architecture has an intrinsic capacity of *perceiving, fitting* and *relating*, that stems from a cultural condition.[15] *Integrating* seems to be the transversal word of a lineage that follows an introspective path, adept of a continuum guiding wire that reinforces the concept of *classic*. Order is implemented in diverse levels (idea, process, drawing, construction, interdisciplinary relationship), through which bridges are established. It is in the cohesion, persistence and solidity of founding values that architecture gains a timeless dimension; it is only on this basis that it is possible to admit ruptures and reinterpretations, because here they are sedimented in a conscious process of fighting against the complexity of the real (physical and social). Towards the generalized crisis scenario and given the relevance of the concept of proportion here demonstrated, it is urgent to redeem (for the theory) a theme that has been forgotten, kept in the past, but still with so much to offer to the practice of architecture: in the face of inescapable diversity, what remains is the power of relations.

References

1. Mangado, F.: Introdução por Francisco Mangado. Arquitectura Ibérica **25**, 4–5 (2008)
2. de la Maza, R.M.: La revolución tranquila. TC-Cuadernos—Serie Dedalo. **57**, 6–7 (2005)
3. Lopes, C.N.L.: João Álvaro Rocha em detalhe. In: Lopes, C.N.L. (ed.). Arquitectura e modos de habitar – Conversas com arquitectos: João Álvaro Rocha, pp. 9–15. CIAMH, Porto (2012)
4. de la Maza, R.M.: Habitar la precisión y otros argumentos cruzados. TC-Cuadernos—Serie Dedalo. **102|103**, 8–15 (2012)
5. Rocha, J.Á.: *Projecto: notas e imagens para um percurso – Trabalho de síntese e relatório de uma aula teórico-prática elaborados para prestação de provas de aptidão pedagógica e capacidade científica.* [Prova de aptidão pedagógica não publicado acessível na Biblioteca do Departamento de Arquitectura da Universidade de Coimbra, não publicado], Porto, Portugal (1996)
6. Rocha, J.Á., Moura, E.S.: João Álvaro Rocha em conversa com Eduardo Souto de Moura. *AI*—Arquitectura Ibérica, **25**, 148–167 (2008)
7. Rocha, J.Á., Maia, J.: *João Álvaro Rocha entrevistado por Joana Maia.* [Gravação áudio para efeitos de tese de doutoramento em curso, com o título provisório "A dimensão intrínseca – Validade do conceito de proporção na arquitetura portuguesa a partir de meados do século XX", não editada], Maia, Portugal (2013, 22 Outubro)
8. Ravalli, A.: In equilibrium upon contemporaneity. In: Caca, Francesco (ed.) João Álvaro Rocha: Architectures, 1988-2001, pp. 6–7. Skira, Milão (2003)

---

[15]One can see the miscegenation capacity of the Portuguese colonization, which, although aggressive, was less authoritarian than the Spanish colonization, according to the architect João Álvaro Rocha [7].

Joana Maia is an Architect by the Artistic College of Porto (ESAP); Master in Methodologies for intervention in the architectural heritage by Faculty of Architecture, University of Porto (FAUP); and has a Degree of Advanced Studies in Architectural and Urban Culture by Department of Architecture, University of Coimbra (Darq FCTUC). Currently developing a PhD thesis under the provisional title *The intrinsic dimension: the value of the concept of proportion in Portuguese architecture from the mid-twentieth century*, at the Department of Architecture, University of Coimbra (Darq FCTUC). Joana Maia teaches Geometry as assistant guest at the Department of Architecture, University of Coimbra (Darq FCTUC), Portugal.

Vítor Murtinho is an Architect and Full Professor at the Department of Architecture, University of Coimbra (UC), Portugal. Since 1988, he teaches the areas of Geometry and Theory and History of Architecture in the Degree Course, Integrated Master's Programme and 3rd Cycles Studies in Architecture. He is senior researcher at the Center for Social Studies of the UC and has interests in the domain of Renaissance theory, architectonics of form and geometry. He was Vice-Rector (March 2011–February 2019) of the UC with responsibility for the heritage, buildings and sustainability. He is the author or co-author of more than one hundred publications (books, book chapters, papers in scientific journals and in conference proceedings). He has participated in several Aproged's conferences. Detailed publications in: http://ces.uc.pt/en/ces/pessoas/investiga doras-es/vitor-murtinho/publicacoes.

Developable Ruled Surfaces from a Cylindrical Helix and Their Applications as Architectural Surfaces

Andrés Martín-Pastor⬤ and Alicia López-Martínez

Abstract The study of developable surfaces has not been very common in the context of architectural education. This is not due to their complexity, but perhaps to the relatively recent emergence of digital tools that enable these surfaces to be controlled via advanced graphic thinking. In our recent workshops on Geometry and Digital Fabrication, we have worked with developable helical surfaces. These workshops have involved the design, manufacture and assembly of two ephemeral pavilions, the *Butterfly Gallery* and the *Molusco Pavilion*. These two experimental structures provided the initial inspiration to enter into the geometrical depth of the expandable conditions of certain helical structures described herein.

Keywords Developable surfaces · Helicoid · Digital fabrication · Ephemeral architecture

1 Developable Helical Surfaces

The study of developable surfaces is strongly marked by the mathematical advances of the seventeenth century and by the figure of Monge [1] who devoted much of his work to studying them in depth. It can be seen that the advance in the knowledge of these surfaces is closely related to spatial intuition and the treatment of space as defined by Monge through his *Géométrie Descriptive* [2]. However, according to Glaeser [3], the presence of such surfaces is not yet common in the context of architectural education and professional practice. This is not due to their complexity, but perhaps to the relatively recent emergence of digital tools that enable these surfaces to be controlled via advanced graphic thinking.

A. Martín-Pastor (✉) · A. López-Martínez
Universidad de Sevilla, Seville, Spain
e-mail: archiamp@us.es

A. López-Martínez
e-mail: alicia.lpmt@gmail.com

© Springer Nature Switzerland AG 2020
V. Viana et al. (eds.), *Thinking, Drawing, Modelling*,
Springer Proceedings in Mathematics & Statistics 326,
https://doi.org/10.1007/978-3-030-46804-0_8

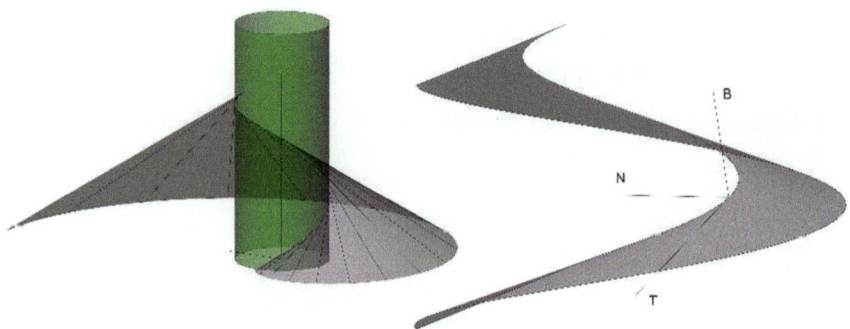

Fig. 1 Left: The helicoid as a surface of equal slope supported on a helix. Right: The tangential developable surface generated by tangent vectors T along a curve. Source: The authors

Specifically, the helicoid is extensively studied in classic manuals of Descriptive Geometry, and its formulation has been described as a surface of equal slope resting on a cylindrical helix [4–7]. We make a description centred on the more general conception of developable surfaces, whereby we consider it as a type of tangential surface with respect to a space curve (Fig. 1) [8]. These tangential surfaces are determined as the envelope surface of the osculating planes [9] (defined by the tangent vector T and the normal vector N of each point of the space curve, the so-called edge of regression). It is common to observe a simplified definition stating that this surface is generated, not by the envelope of osculating planes, but directly by the set of tangent vectors T along the edge of regression, as mentioned by Glaeser [3, p. 63]. According to this definition, and within the general theory of surfaces, the developable surface from a cylindrical helix can be understood as a tangential developable (i.e. torsal) ruled surface of this curve.

2   Evolute, Involute and Evolvent

On the other hand, we can understand the helicoid as a surface of equal slope that rests on a flat horizontal curve called involute of the circumference or 'evolvent' (according to the Spanish notation). These concepts are reviewed here.

In general, the evolute is the curve formed by the orbit of the centres of all osculating circles of another curve, called the involute. A single evolute has infinitely many equidistant involutes (Fig. 2, left) and the shape of the evolvent is identical for all circumferences, according to Elizalde [9, p. 334]. This may seem contradictory to the aforementioned property: a single evolute (circumference) has infinite involutes (evolvents) but, as shown in Fig. 2 (left), the infinite evolvents would be interpreted as different rotations of the curve around the centre of the circumference. The turns produce equidistant involutes and these become into the original in a complete rotation. Hence, the division of the circumference into n equal parts is directly

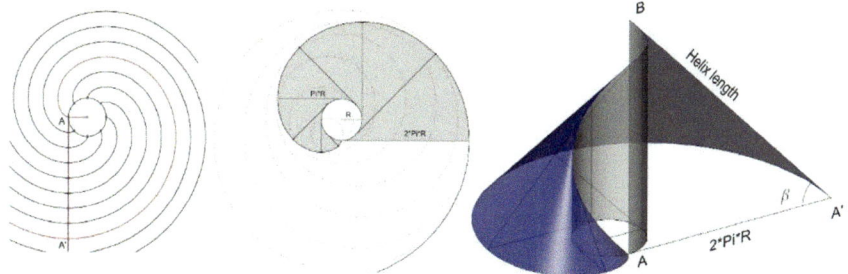

Fig. 2 Left: Infinite evolvents of the circumference. Centre: relationship between the length of the circumference and its evolvent. Right: Graphic demonstration. Source: The authors

related to the division of segment *A*-*A'* (tangent to the circumference) into the same *n* equal parts. The evolvent also maintains another important property with respect to its circumference: The distance between the start point *A* and the end of its first rotation *A'* is strictly $2\pi R$, the length of the circumference. This relationship can be extrapolated to any proportional distance (Fig. 2, centre).

The helix has a constant slope β, just like the straight lines of the helicoid. If we unroll the vertical cylinder that carries the helix *A*-*B*, it becomes a segment *A'*-*B*, since that cylinder is the rectifying surface of the helix [8, p. 72]. The rectified segment *A'*-*B* must be the length of the helix and segment *A*-*A'* must be the length of the circumference. We can graphically verify this spatial correspondence between the unrolled cylinder and the helicoid (Fig. 2, right). The rectified segment *A'*-*B* coincides with the generating lines of the helicoid and the evolvent passes through point *A'*.

3   Factors of the Development

3.1   *Development by Circular Sectors*

If the helicoid is formed by generating lines of equal length, as in Fig. 3, then its planar development is a portion of a circular sector limited by two lines tangent to the smaller circumference [9, p. 338]. The length of the arc of this smaller circumference is equivalent to the length of the cylindrical helix on which the straight lines are tangentially supported.

The curvature at each point of the generating helix is defined by the osculating circle, and the radius of curvature *OA* coincides with the radius of the minor circumference in the development. The points *OB*, of the right-hand side triangle contained in the osculating plane passing through point *A*, determines the largest circumference (Fig. 3). We can imagine how the planar development curves in space until it attains

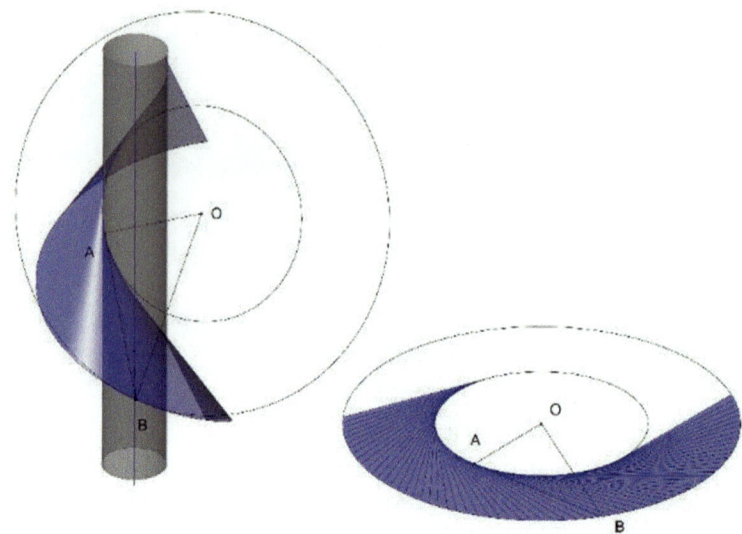

Fig. 3 Relationship between the osculating circle in space and the planar development. Source: The authors

the helicoidal form (Fig. 3 left). Therefore, the same planar development could be considered common to several helicoids.

3.2 Developments of Helicoids Generated by Evolvents

If the helicoid is generated as an equal slope surface that rests on an evolvent, then it can be verified that the unrolled surface is another evolvent (Fig. 4). The first evolvent comes from the circle projection of the helix, and the second from the osculating circle of the helix. The most important property, which we will use later, is that both evolvents, and their respective circles, have a homothetic relationship.

4 Curvature of a Curve and Curvature of a Surface

The main curvature δ of a curve (or simply, curvature) measures the angular deviation between two normal vectors N, infinitely close to a point on the curve. This curvature is characterized graphically by the osculating circle, which is defined by three points of the curve that are infinitely close to each other.

In contrast, the torsion ζ of a curve evaluates the tendency of this curve to rise from its osculating plane and it is measured by the angular deviation between two

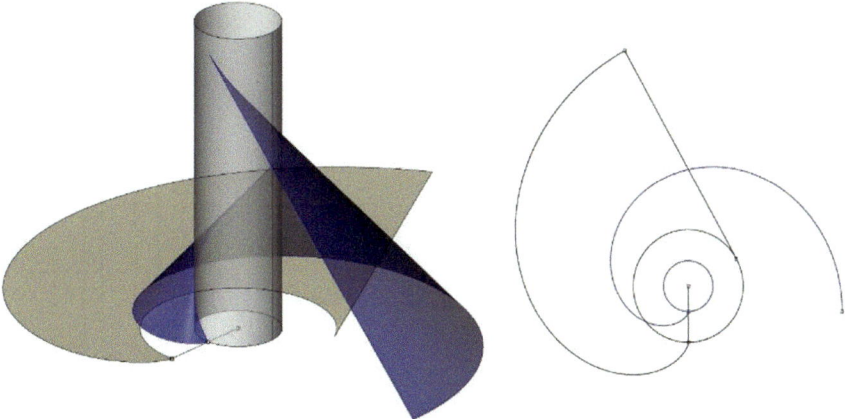

Fig. 4 Left: Relationship between the evolvent of the basis and the unrolled helicoid. Right: Homothetic relation between both evolvents. Source: The authors

binormal vectors B, that are infinitely close. Once the curve unfolds on a plane, it maintains the main curvature but it loses the torsion, according to Leroy [8, p. 143].

The evolution of the values of δ and ζ can be verified by their equations, by means of substituting values:

$$\text{Curvature } \delta = \frac{R}{R^2 + K^2} \qquad \text{Torsion } \zeta = \frac{K}{R^2 + K^2}$$

$$x(t) = R \cos t \quad y(t) = R \sin t \quad x(t) = K t$$

The curvature of a surface, in general terms, is characterized by two principal curvatures determined by two principal directions V, U. In the case of the developable surfaces, one of these two curvatures is always zero, since it coincides with the direction of the generating lines. For this reason, we usually denominate the developable surfaces as surfaces of a single curvature, which is equivalent to having a vanishing Gaussian curvature, or as formed only by parabolic points (see Glaeser [3, p. 61]).

This has major practical consequences: the curvature of these surfaces is characterized in each point by a single parameter of curvature associated perpendicularly to the direction of the generating lines. It can, therefore, be expressed graphically as a circle of curvature, as seen in Fig. 5.

If we relate these geometrical aspects to the possibility of constructing these surfaces with laminar materials, such as wood and veneers, then this curvature value can help us determine the feasibility of using such materials. The appropriate use of these materials would be limited between an infinite radius of curvature (zero curvature) and a minimum radius (maximum curvature of the material) [10]. To conclude this first approach to design, we use another type of analysis where tensions and deformations are considered.

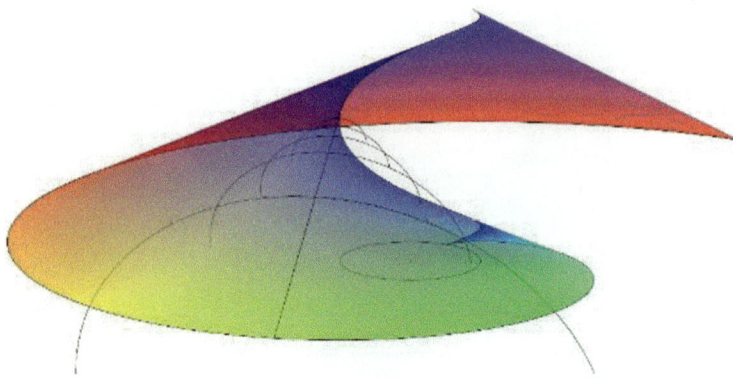

Fig. 5 Curvature of a developable surface characterized at each point by unique parameters (circle of curvature) perpendicular to each generating line. Source: The authors

5   Helicoids that Share the Same Development

Once the geometrical foundations are reviewed, we establish that several helicoids can share the same planar development, since this property explains why they can be turned into transformable structures.

We have formulated a parametric algorithm—based on tangents T along the regression edge—employing Rhinoceros[TM] and Grasshopper[TM] software. We have been able to relate how the flat development of a helicoid varies by changing, or keeping constant, different parameters (radius R of the helix, pitch of the helix, radius of the osculating circle …) as we will show below.

5.1   Helicoids that Share the Same Planar Development

The same planar development offers an infinite number of possible helicoids since, in fact, there are two parameters to combine. First, the radius Rc of the cylinder is limited between zero and the smallest circle of the development. Secondly, we can vary the pitch of the helix for each Rc chosen. The pitch would be comprised between two limits: zero, at which the helicoid acquires the form of a cone coiled on the cylinder; and a maximum pitch position for each Rc.

In Fig. 6 (left), it can be observed how the planar development can be adjusted around the same cardboard cylinder $R1$ in different helicoids, varying from zero pitch (cone), to the maximum pitch position. In the graphic study of Fig. 6 (right), we verify that the generating lines of the helicoid are tangent to another invisible inner cylinder that, in the case of maximum pitch, coincides with the cylinder of radius $R1$.

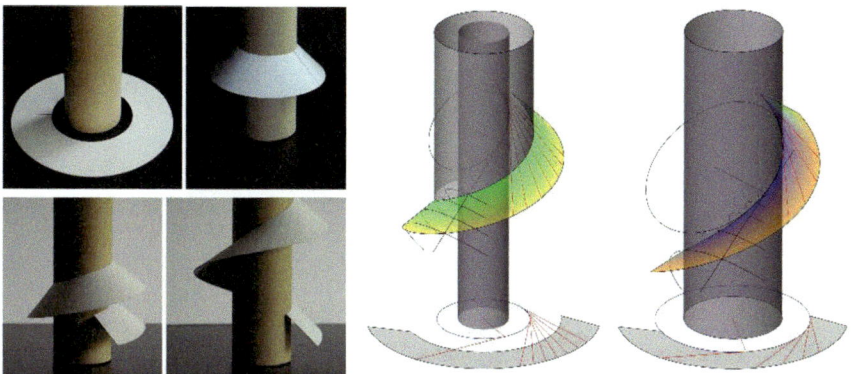

Fig. 6 Left: Different helicoids with similar development, around the same cylinder. Right: Evolution of the generating lines to maximum position (rightmost figure). The maximum curvature occurs in the proximity of the helix, where the lines are tangent. Source: The authors

5.2 Family of Helicoids that Share the Same Development (Circular Sector) and Whose Generating Lines Are Tangent to the Helix

In order to study how a series of helicoids share the same development, we keep constant the length of the helix and the radius of the osculating circle, while we vary the radius of the helix (Fig. 7). This radius thus changes from the maximum value (coincident with the radius of the osculating circle in planar position) to a value close to zero. As the radius of the helix is reduced, it also decreases the pitch of the helix. The helix must maintain the same length; consequently, the helicoid begins to roll up. In the limit position (radius close to zero), the helix tends to become a vertical segment, since it has been coiled almost an infinite number of times.

5.3 Development by Evolvents

Similarly, when the helicoids are generated by evolvents, the same planar development is common to several helicoids, which guarantees its transformable character as shown in Fig. 8. It can be observed how the radius of the helices decreases homothetically with the evolvent (Fig. 8, top). From another point of view, we can deduce that all the helicoids supported by the same evolvent, regardless of the slope of the helical surface have, as edge of regression, helixes of equal radius R (Fig. 8, bottom-left). So, we conclude that we can always create a transformable structure from any two helicoidal developments, joining the evolvents and providing that both curves are

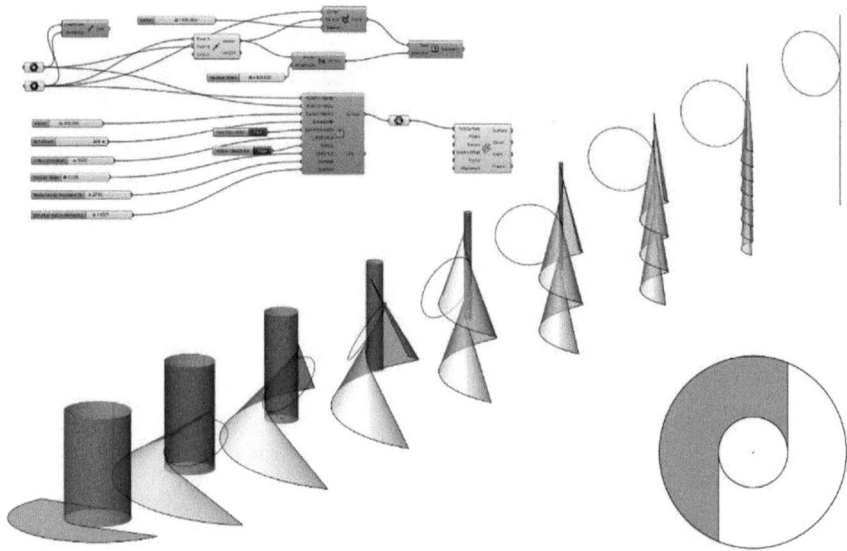

Fig. 7 Family of helicoids that share the same planar development and whose generating lines are tangent to an interior helix of variable radius. It can be appreciated how all the generating helices share the same osculating circle radius. Source: The authors

of equal length. The two helicoids that share the evolvent will have two generating helixes of equal radius at each position (Fig. 8, bottom-centre).

A unique position of this transformable structure formed by two helicoids occurs when one of them is plane, denoting the minimum angle of the other helicoid. If a third helicoid is added, then we arrive at one formal solution of great structural resistance (Fig. 8, bottom right). As Fig. 8 demonstrates, in addition to the explanation, the helicoids that share the same evolvent are related to a spatial affinity transformation.

6   Transformable Helicoidal Folded Structures

We now propose a set of transformable helicoids, formed by plane surfaces sewn at their edges.

6.1   Folding on Concentric Rings

We propose a series of circular sectors with their edges sewn at their contact line, which allows the movement of the whole set. We will consider the physical movement involved in carrying the folded geometric structure from the plane to space (Fig. 9).

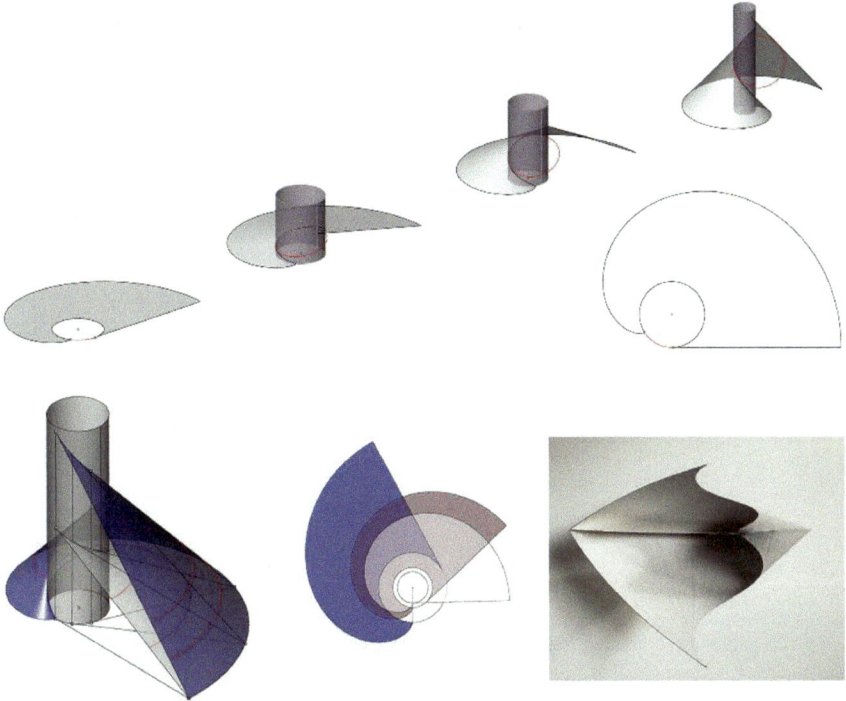

Fig. 8 Top: Family of helicoids that share the same planar development and whose generating lines are tangent to helices. Bottom left: Two helicoids supporting the same evolvent always share the radii of their generating helixes Source: The authors

We created a model to study different positions in its extensible movement and to inquire which geometric relationships link the helicoid family with the motion of the set (Fig. 9). We first deduced that all the rings of the set are helicoids of equal pitch. It is important to highlight how contact occurs between generating lines and their respective apparent helices. Only a truly tangential contact occurs in the case of the smaller rings, the rest of the ruled lines of the set are not tangent to the apparent helices, but to other inner helices (Fig. 10). These helices have a radius similar to the smallest circumference of the circular sector. It can, therefore, be deduced that the whole set depends on a unique generator helix, and all the helical rings are fragments of one or the other branch of the same helicoid in different rotations.

As we have already observed, the generating lines of the helicoid are tangent to the helix only when the maximum pitch position is reached, and hence the helicoid corresponding to the minor ring is always in the maximum pitch position (Fig. 10, centre). If we keep the pitch invariant, then we diminish the degree of freedom and the expandable structure has only one possible transformation path. With more tension exerted, and the maximum pitch mode achieved, the helicoid is deformed only by

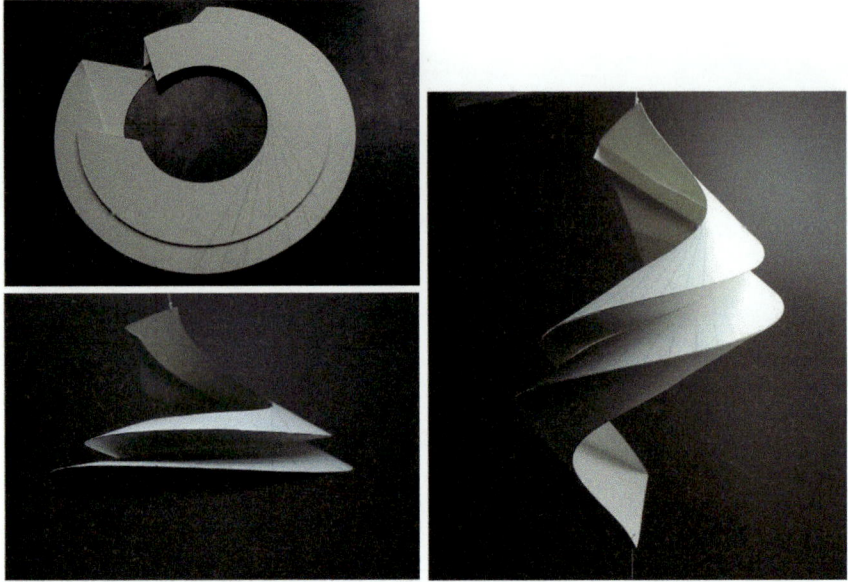

Fig. 9 Set of circular sectors, connected at their common edge. Evolution from their folded and plane position to various helical rings. Source: The authors

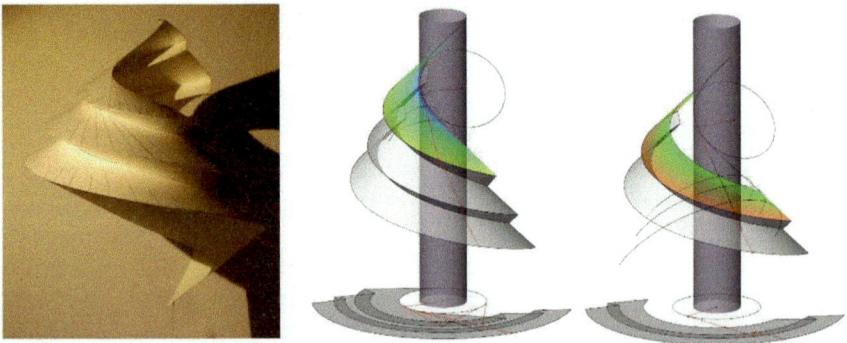

Fig. 10 Left: Structure formed by a group of helical rings. Centre and right: The ruled lines of the second ring are tangents to the cylinder of the minor ring. The maximum curvature of surface occurs in the proximity of the minor ring helix. Source: The authors

vertical lengthening, while its radius is reduced. In this way, in its motion, the whole transformable structure depends solely on the smaller ring.

6.2 Folding Evolvents

Another possibility of folding consists of articulating the helicoids with respect to a mirror plane, which imposes the condition of equal slope to that plane. Since the helicoid is a surface generated by a family of ruled lines of equal slope with respect to the plane of the evolvent, the set of lines could be reflected specularly from this plane, thereby retaining the same slope in both cases (Fig. 11). The evolvent can, therefore, be understood as the folding line of a single continuous surface (Fig. 11 bottom).

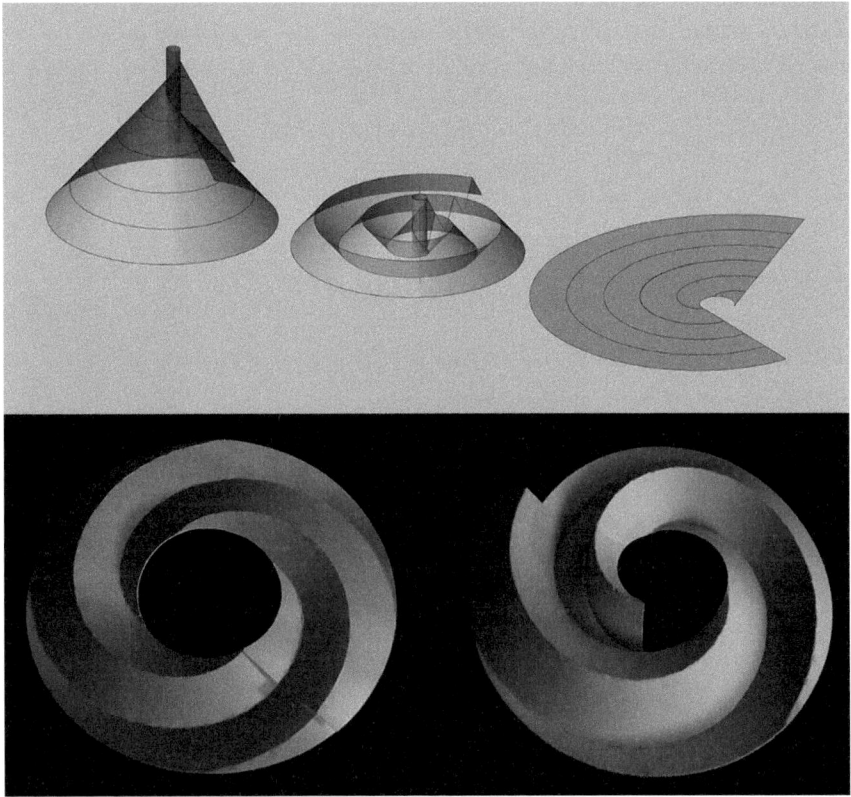

Fig. 11 Top: Sections (evolvents) of a helicoid using horizontal planes. Bottom: A single piece of cardboard folded along evolvents. Source: The authors

7 Conclusions and Discussion

We have analysed the generation of helicoids from their parameterization in Rhinoceros–Grasshopper and paid special attention to developments in circular rings and evolvents. We are now able to propose structures formed by several helical surfaces linked together, giving rise to a single transformable or extensible structure. These surfaces can be built with laminar pieces, which can be combined with different folded structures, thereby providing a wide variety of formal solutions. We have studied the geometric relationships between the helicoids in the movement of transformation of a linked set and have demonstrated the importance of certain geometric parameters over others. All this expertise and knowledge can be put into practice when we design helical surfaces for use in lightweight architecture. At this point, we have designed two architectural installations on the general properties of the helicoids (Figs. 12 and 13) [11]. The next challenge is now to design an extensible structure that will benefit from the characteristics studied.

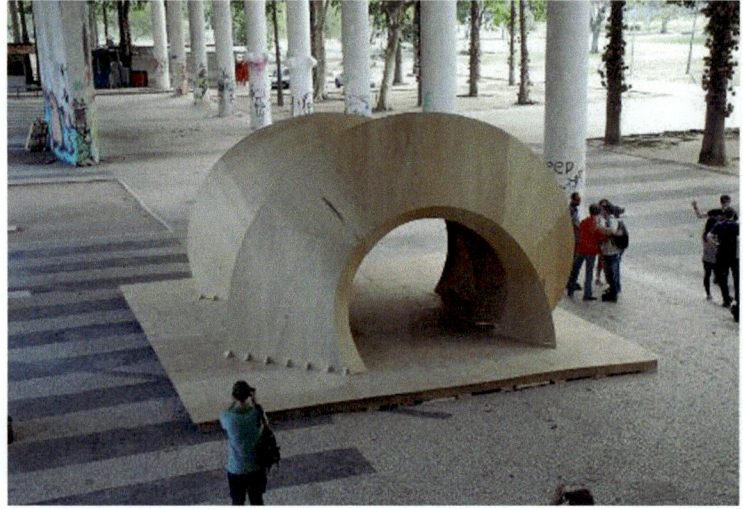

Fig. 12 The Butterfly Gallery, Universidade Federal do Rio de Janeiro, Brazil, 4–14 August 2015. Source: The authors

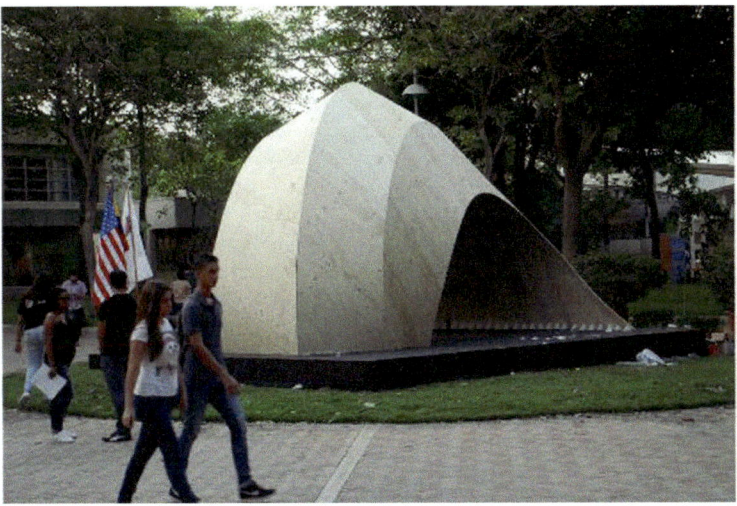

Fig. 13 The Molusco Pavilion. Facultad de Arquitectura, Urbanismo y Diseño. Universidad del Norte, Barranquilla, Colombia. October 2016. Source: The authors

References

1. Monge, G.: Sur les développées des courbes à double courbure et leurs inflexions. Journal encyclopédique, 284–287 (1769)
2. Monge, G.: Géométrie Descriptive, 1st ed. in e Les Séances des écoles normales recueillies par des sténographes et revues par des professeurs, Paris (1795)
3. Glaeser, G., Gruber, F.: Developable surfaces in contemporary architecture. J. Math. Arts **1**(1), 59–71 (2007). https://doi.org/10.1080/17513470701230004
4. Izquierdo Asensi, F.: Geometría descriptiva superior y aplicada. Dossat, Madrid (1986)
5. Gentil Baldrich, J.M.: Método y aplicación de representación acotada y del terreno. Bellisco (1998)
6. Rodrigues, A.J.: Geometría Descritiva, vol. 2. Imprenta Nacional, Río de Janeiro (1945)
7. Taibo Fernández, A.: Geometría descriptiva y sus aplicaciones, vol. 2. Tébar Flores, Madrid (1983)
8. Leroy, C.F.A.: Traité de Geométrie Descriptive, 10th edn. Gauthier Villars, Paris (1877)
9. Elizalde, J.A.: Curso de Geometría Descriptiva [& Atlas]. Manuel Tello, Madrid (1882)
10. Narváez-Rodríguez, R., Martin-Pastor, A., Aguilar-Alejandre, M.: The caterpillar gallery: quadric surface theorems, Parametric design and digital fabrication. In: Advances in Architectural Geometry 2014, pp. 309–322. Springer (2014). https://doi.org/10.1007/978-3-319-11418-7_20
11. Martín-Pastor, A.: Um retorno aos fundamentos da geometría. The Butterfly Gallery, Estratégias Geométricas para a Fabricação Digital. Cadernos PROARQ. Revista de Arquitetura e Urbanismo **25**, 2–30 (2016). Retrieved from http://cadernos.proarq.fau.ufrj.br/en/paginas/edicao/25

Andrés Martín-Pastor is an Architect with a Ph.D. from the Universidad de Sevilla (Spain), where he works as Lecturer in the Department of Graphic Engineering. His research is focused on perspective and geometry in architecture and makes a thorough review ranging from the inherited graphic tradition until today's digital tools. His research also includes the study of developable surfaces and their applications in lightweight architecture. In 2015, he received the Emporia Silver Award Innovation in Ephemeral Architecture. He has lectured at several International Universities and has published numerous books, chapters and articles.

Alicia López-Martínez holds degrees in Building Engineering and a master degree in Engineering. She is a Ph.D. student at the Universidad de Sevilla (Spain). Her main research interest lies in the application of developable surfaces and digital fabrication in architecture.

Porous Spatial Concrete Structures Generated Using Frozen Sand Formwork

Hannah Müller, Christoph Nething, Anja Schalk, Daria Kovaleva, Oliver Gericke, and Werner Sobek

Abstract One of the most distinctive properties of concrete is its ability to take nearly any shape while being processed in a liquid state and retain that shape upon hardening. Thus, the variety and complexity of concrete structures are virtually unlimited. However, nowadays concrete continues to be used mainly for simple volumetric objects, with the geometrical complexity being mainly constrained by the design and production of the formwork. In order to allow for more complex geometries, new formwork methods should be developed. In the framework of a research project on formwork methods for concrete, conducted at the Institute for Lightweight Structures and Conceptual Design (ILEK) at the University of Stuttgart, a new method called *Hydroplotting* was developed and tested. The present study provides a general description of the method and discusses its potential for the fabrication of concrete objects with a high geometrical complexity.

Keywords Concrete structures · Digital fabrication · Material-based design · Complex geometries

H. Müller (✉) · C. Nething · A. Schalk · D. Kovaleva · O. Gericke · W. Sobek
Institute for Lightweight Structures and Conceptual Design (ILEK), University of Stuttgart, Stuttgart, Germany
e-mail: hannahlauramuller@gmail.com

C. Nething
e-mail: christoph.nething@ilek.uni-stuttgart.de

A. Schalk
e-mail: anja.schalk@web.de

D. Kovaleva
e-mail: daria.kovaleva@ilek.uni-stuttgart.de

O. Gericke
e-mail: oliver.gericke@ilek.uni-stuttgart.de

W. Sobek
e-mail: werner.sobek@ilek.uni-stuttgart.de

© Springer Nature Switzerland AG 2020
V. Viana et al. (eds.), *Thinking, Drawing, Modelling*,
Springer Proceedings in Mathematics & Statistics 326,
https://doi.org/10.1007/978-3-030-46804-0_9

1 Concept and Approach

The focus of the research presented in this paper was on the development of a formwork method that overcomes the constraints of conventional methods in terms of geometrical complexity [1], and that, at the same time, is also relevant in economic and ecological terms [2].

Previous research in this field has shown that granular materials such as earth [3], clay or sand [4] can be used as formwork materials for concrete. These materials can be sculpted into the desired shape, and they can also be easily removed from a hardened concrete piece. Thus, they allow for the creation of geometrically complex features, such as spatial networks or bottlenecks. Various binders can be admixed to these granular materials to ensure their geometrical stability during the casting and hardening of the concrete. For example, the admixture of water to sand results in an increase of geometrical stability. This effect can be significantly amplified when the mixture is frozen [5]. Using numerically controlled machining methods, such as CNC-mills, this mixture can be sculpted into high-precision complex formworks.

The sculpting of frozen sand allows for complex geometries of the formwork, but there are still certain constraints—the most important one being the fact that the concrete objects thus obtained are still only volumetric (even if of a higher geometric complexity). In order to overcome these constraints and to allow for the creation of highly complex spatial structures in concrete, ILEK developed the formwork method *Hydroplotting* in the framework of an interdisciplinary research project.

2 Method

Hydroplotting builds on the basic principle of formworks made from frozen sand, where the freezing of moistened sand leads to a stable formwork geometry (Fig. 1). However, instead of mixing all the sand with water at the same time, *Hydroplotting* uses a selective injection of water into predefined sections of a given volume of dry sand, where the water seeps into a limited space around the injection point. When the sand is frozen, the wet parts turn into a solid state, while the dry sections still exhibit granular material properties and can easily be removed, leaving a spatial network.

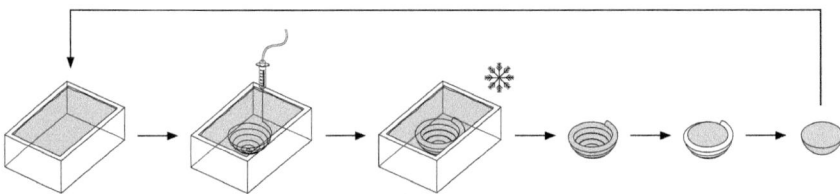

Fig. 1 *Hydroplotting* process

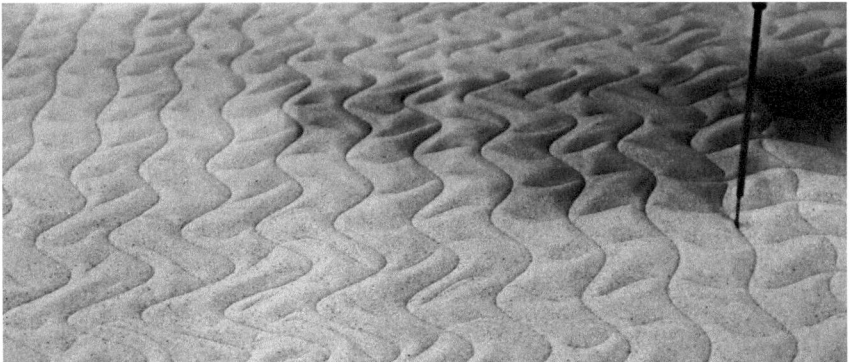

Fig. 2 Needle trajectory in the sand

The water is injected by a custom-made dripping device, installed on a CNC-machine. The dripping device consists of a water tank connected to a needle, the movement of which is numerically controlled (Fig. 2). The geometrical properties of the resulting formwork are defined by the amount of water per second (feed-rate), as well as by the velocity and the trajectory of the moving needle. The geometrical configuration of the trails of water-saturated sand may vary from spheres to cylinders, due to capillary effects. Even cone shapes are possible by intended variation of the needle's velocity along the injection path.

The path for the CNC-guided needle is created in a digital design environ-ment. Unlike traditional CAD-CAM transfer processes (where the geometry of an object designed in CAD software is subsequently subjected to automated trajectory-planning routines in CAM software), *Hydroplotting* generates the path of the needle directly in the design environment. Fabrication parameters such as velocity or feed-rate are integrated as design variables that determine future geometrical properties. When the pattern is generated, the information is forwarded as G-code to the CNC-machine. There it is plotted, with the use of a CNC-guided needle, by the injection of water into the dry volume of sand. Due to its low density difference from the dry sand, the saturated sand preserves its position within the dry sand until the freezing process is finalised. During freezing, moistened parts of the body of sand are trans-formed into a solid state, whereas the dry sand keeps its granular consistency. Once the volume of sand is completely frozen, the dry sand can easily be removed.

After the dry sand has been removed, the leftover frozen sand represents the "negative shape" of the intended concrete piece (Fig. 3). During the curing of the concrete, the formwork is put into ambient conditions allowing for the sand to thaw. Upon appropriate curing of the concrete, the formwork material is removed from the concrete piece.

The underlying principle of the process combines the inherent material properties of sand and water with the controllability of digital manufacturing. However, in con-trast to additive manufacturing methods, as 3D printing [6] or subtractive methods,

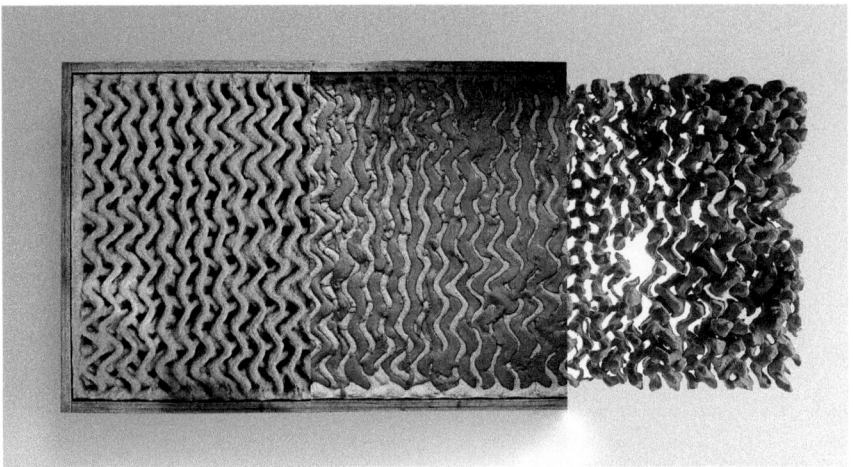

Fig. 3 Photomontage of different stages in concrete piece production via *Hydroplotting*. From left to right: Frozen sand formwork, cast concrete in formwork, final concrete piece

such as CNC-milling [5], *Hydroplotting* is not based on the translation of the geometrical characteristics of a digitally designed object into a manufacturing process. It rather integrates material and fabrication parameters into the design environment. These parameters are first, the distribution of the water inside the sand in relation to the feed-rate; second, the change of the shape of the infiltrated water as a function of the velocity of the needle; and last, the effect of fusion of two injected parts in close proximity to each other. Thus, the geometrical characteristics of the final object are influenced both by material and process parameters of the formwork (Fig. 4). On the one hand, these factors may seem to restrict the freedom of design; on the other hand, these parameters and constraints define conditions for a virtually infinite amount of design options in the framework of the fabrication method that are all producible per se.

3 Prototype

The method was tested and improved on a series of prototypes. Complexity was gradually increased, beginning with an investigation of basic process parameters and ending with the design and fabrication of complex spatial structures offering specific geometrical features. For the first prototype, the influence of process parameters, such as the feed-rate, was explored using a punctual injection with a steady increase in the amount of injected water. This resulted in spheres of saturated sand with gradually increasing diameters. The resulting concrete structure revealed a viable gradient in porosity (Fig. 5). In a second step, the type of pattern was exchanged from punctual to linear, where the variation of the velocity of needle movement led to a variation in

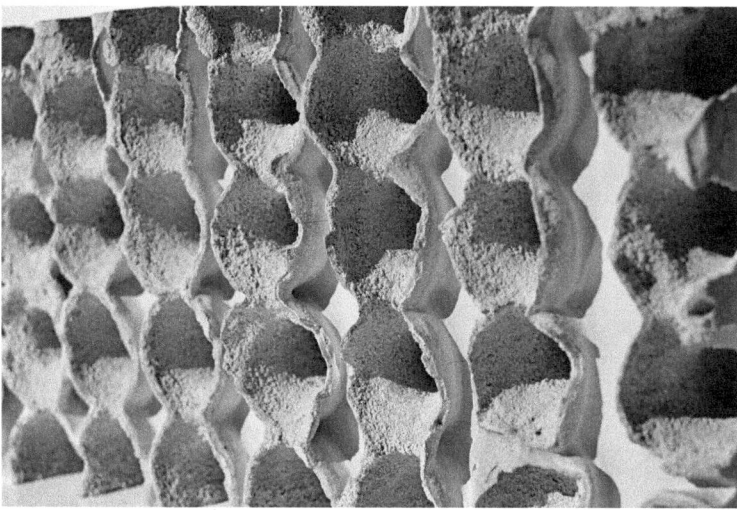

Fig. 4 Influence of sand formwork on concrete surface

the diameter of sand struts, and consequently, in the density of the resulting concrete structure (Fig. 6). Moreover, the character of the formwork's geometry (punctual or linear patterns (Figs. 5 and 6)) influenced the character of interconnectivity within the concrete structure.

In further tests, the line-based type of trajectories was extended to curves with a variable curvature, for example, sine curves with variable amplitudes. These variations caused a gradient in the density of the frozen sand network which was directly inherited by the resulting concrete structure (Fig. 7). Due to the interdependence between the geometrical properties of base curves, their variations and the distribution of water throughout the sand, the resulting geometries appeared as complex spatial structures, even when based on simple patterns.

Currently, the *Hydroplotting* method is being improved with regard to the quality and precision, as well as to the scalability for the production of large-scale concrete structures and further spatial potentials.

4   Conclusion

Based on the analysis of the results achieved so far, *Hydroplotting* seems to be a very promising method for creating formworks, particularly in regard to the geometrical complexity and potential for scalability. By investigating more complex trajectory-planning strategies and refining the interactions of parameters, the method can be made applicable for the creation of architectural objects on a larger scale, offering unique geometrical, visual and haptic properties. The authors believe that

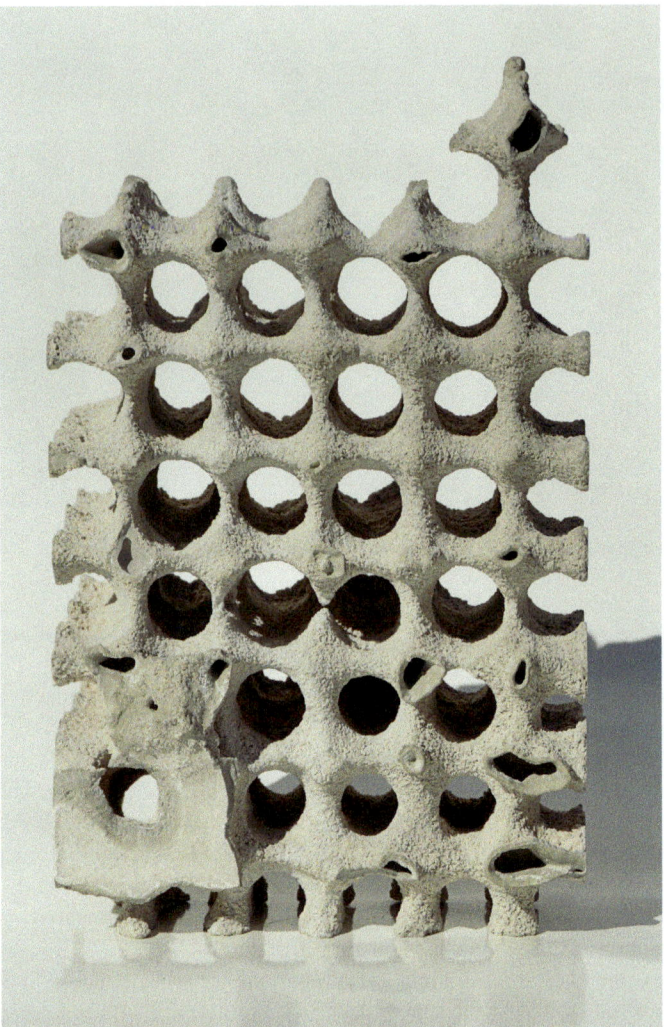

Fig. 5 Prototype of concrete structure based on point grid

this method is highly relevant for architectural design, allowing, for example, for a hitherto impossible translucency of concrete structures or an opening of the borders between material and void. In conclusion, *Hydroplotting* may contribute to accomplish new qualities in the design perception of architectural space.

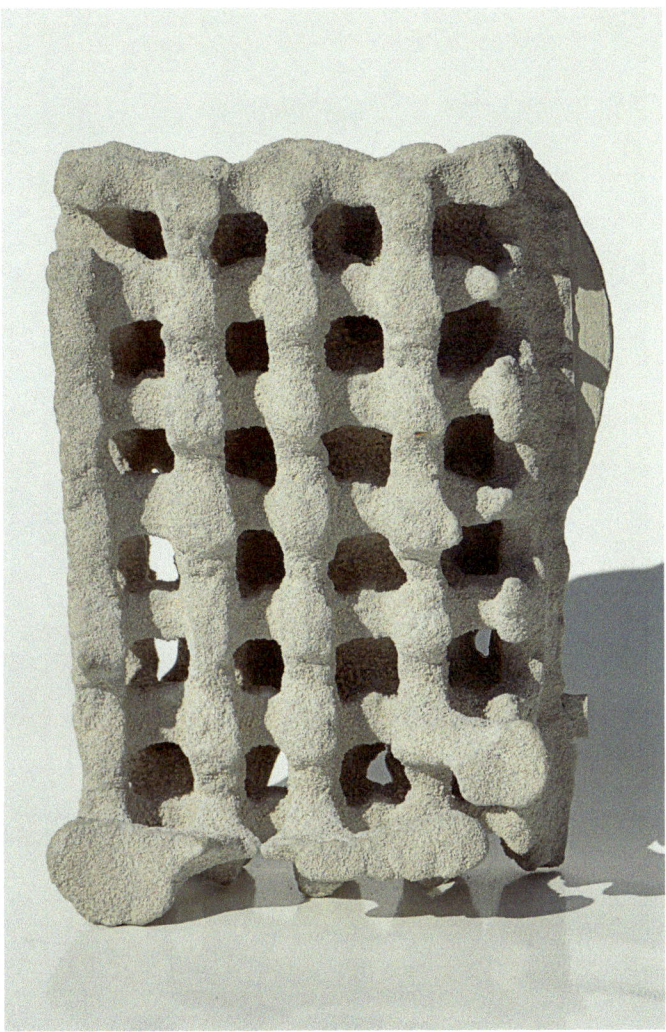

Fig. 6 Prototype of concrete structure based on linear grid

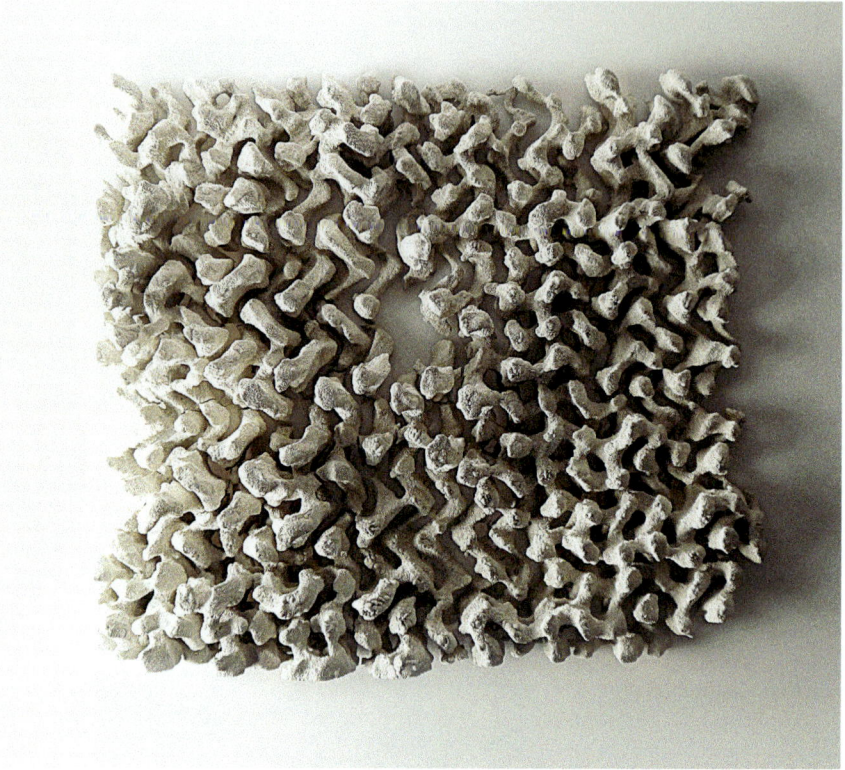

Fig. 7 Concrete structure prototype derived from sine curves with varying amplitudes

Acknowledgements We gratefully acknowledge Holcim GmbH and Sika AG for their material donations and support.

References

1. Cutler, B., Whiting, E.: Constrained planar remeshing for architecture. In: Proceedings of Graphics Interface 2007, Montréal, Canada (2007)
2. Graham, P.: Building Ecology: First Principles for a Sustainable Built Environment. Blackwell Science, Oxford, UK (2003)
3. Isler, H.: New shapes for shells. Bull. Int. Assoc. Shell Struct. **8**, 123–130 (1961)
4. Gericke, O., Haase, W., Sobek, W.: Schalungsmethode zur nachhaltigen Herstellung von Beton-bauteilen mit gekrümmten und unstetigen Oberflächen. In: Leicht Bauen mit Beton - Forschung im Schwerpunktprogramm 1542, Förderphase 1, 1, pp. 218–223. Institut für Massivbau, Technische Universität Dresden, Dresden (2014)
5. Gericke, O., Kovaleva, D., Haase, W., Sobek, W.: Fabrication of concrete parts using a frozen sand formwork. In: Kawaguchi, K., Ohsaki, M., Takeuchi, T. (eds.) Proceedings of the IASS Annual Symposium 2016, Tokyo, Japan (2016)

6. Lim, S., Buswell, R., Le, T., Wackrow, R., Austin, S., Gibb, A., Thorp, T.: Development of a viable concrete printing process. In: Proceedings of the 28th International Symposium on Automation and Robotics in Construction, Seoul, Korea (2011)

Hannah Müller is an architectural graduate currently working with Dietz Joppien Architekten in Frankfurt am Main. In 2018, she completed her master's degree of Architecture and Urban Planning from the University of Stuttgart at the Institute for Lightweight Structures and Conceptual Design (ILEK). During her studies, she worked as a student assistant at ILEK and as a tutor at the Institute of Building Structures and Structural Design (ITKE). Her research interests are material based and structural design, digital fabrication in architecture and lightweight structures.

Christoph Nething is Research Associate at the Institute for Lightweight Structures and Conceptual Design (ILEK) at the University of Stuttgart, Germany. In 2018, he completed his master's degree of Architecture and Urban Planning at the Institute for Lightweight Structures and Conceptual Design (ILEK) on the topic of bio-cementation. His research interests include material-based and structural design, small- and large-scale digital fabrication techniques as well as lightweight structures.

Anja Schalk is Project Engineer at KREBS + KIEFER Ingenieure GmbH in Mannheim, Germany. In 2019, she completed her master's degree of Civil Engineering at the University of Stuttgart's Institute for Construction and Design (KE) with a specialisation in constructive engineering. During her studies, she worked as a tutor at the Institute for Lightweight Structures and Conceptional Design (ILEK) from 2013 to 2015 and as a student assistant at KE. Her research interests are material research and lightweight structures.

Daria Kovaleva is Architect and Research Associate in the Institute for Lightweight Structures and Conceptual Design (ILEK) at the University of Stuttgart, specialising in materially informed design and fabrication strategies. After receiving her Diploma in Architecture from Moscow Architectural Institute in 2009, she worked for various architecture offices and construction companies including Moscow and Stuttgart departments of Werner Sobek Group. Since 2014, she joined ILEK, where she focuses on fabrication of bio-inspired functionally graded concrete structures and their applications in construction industry.

Oliver Gericke is Research Associate at the Institute for Lightweight Structures and Conceptual Design (ILEK) at University of Stuttgart, Germany. His work focuses on the design and fabrication of segmented concrete shell structures. Recently, he was part of the team of researchers that developed the Frozen Sand formwork method for the production of complex shaped concrete structures. In 2013, he was awarded a diploma in Civil Engineering by the University of Stuttgart. Since then, he works at ILEK, where he is also enrolled as a Ph.D. student.

Werner Sobek is Architect and Consulting Engineer. He heads the Institute for Lightweight Structures and Conceptional Design (ILEK) at the University of Stuttgart. From 2008 until 2014, he was also Mies van der Rohe Professor at the Illinois Institute of Technology in Chicago and guest lecturer at numerous universities in Germany and abroad, e.g. in Austria, Singapore and the USA (Harvard). In 1992, Werner Sobek founded the Werner Sobek Group, offering premium consultancy services for architecture, structures, façades and sustainability. The Werner Sobek Group has offices in Stuttgart, Dubai, Frankfurt, Istanbul, London, Moscow and New York.

Calculated Geometries. Experiments in Architectural Education and Research

José Pedro Sousa

Abstract This paper addresses the role of geometry in architecture over history as the language supporting and bridging the design and construction realms. It does so by confronting two different, but complementary, approaches: the *descriptive* and the *generative*. By defending the creative relevance of the latter, this paper examines how the analogical and the digital conditions have supported and stimulated such conceptual and operative interest. Although a generative approach to geometry does not depend on the use of computers, this paper argues that computation is refactoring the role of geometry in architecture by merging the power of calculation with that of representation. To support and illustrate such consideration, five teaching and research works conducted by the author are here presented and illustrated.

Keywords Architectural geometry · Constructive geometry · Computational design · Generative design · Digital fabrication · Robotics · 3D printing

1 Introduction

1.1 On Geometry in Architecture

Over history, geometry has been considered as the language supporting and bridging the design and construction realms. When Mitchell [1] argued that "architects tend to draw what they can build, and build what they can draw", he is indeed placing the mastery of geometry (i.e. drawing) aside with the knowledge on materiality (i.e. construction) as the two domains that, by mutual influence, define the space of architectural possibilities. Therefore, the universe of creative and material solutions that can be imagined, described and built by architects strongly depends on the degree in which their geometries can be controlled.

J. P. Sousa (✉)
Digital Fabrication Lab, Faculdade de Arquitectura, Centro de Estudos de Arquitectura e Urbanismo, Universidade do Porto, Porto, Portugal
e-mail: jsousa@arq.up.pt

© Springer Nature Switzerland AG 2020
V. Viana et al. (eds.), *Thinking, Drawing, Modelling*,
Springer Proceedings in Mathematics & Statistics 326,
https://doi.org/10.1007/978-3-030-46804-0_10

1.2 The Descriptive Versus the Generative Approach

In the exploration of geometry in design, it is possible to distinguish two different, though complementary, approaches. In broad terms, I will call them the *descriptive* and the *generative*. In the first one, geometry is understood as a fixed and metric tool to describe objects and spaces. That is the case, for instance, when an axonometric drawing, or a set of dimensioned orthographic projections, are employed to define a single and unique design. In a different way, when Albrecht Dürer related different facial profiles in a system of adjustable geometric alignments, he was opening the possibility for describing an endless world of facial profile versions [2]. Such approach to geometry can be said as *generative,* and unlike the previous one, it is eminently relational and topological.

2 Calculated Geometries

2.1 The Analogical Condition

Due to their range of creative indeterminacy, one can argue that generative approaches are highly relevant for design thinking and practice as they trigger our imagination in a structured as well as inventive way. Rather than prescribing fixed and explicit design intentions, many architects have been exploring systems of implicit designs, which could lead to the generation of multiple variable solutions. Using drawings, Leon Battista Alberti's system of proportions for designing columns [3], Guarino Guarini's topological studies of solids [4], or Jean-Nicolas-Louis Durand's combinatorial system for building design [5], reveal an interest in the generative dimension of geometry. However, when drawing was a limited medium to imagine and define such generative systems, some architects turned into other physical techniques and means to pursue such aspirations. This was the case of Antoni Gaudí's hanging chain models [6] or Frei Otto's soap-film experiments where physical systems were devised to control the generation of design geometries [7].

Independently of the medium or techniques employed, the scenarios described in the current section are not concerned with defining the geometry of a single design solution but, instead, with establishing the geometric conditions that rule the generation of multiple possible designs. Thus, one can argue that the geometry of such designs is not simply drawn but, instead, is the product of some sort of calculation procedure. For instance, in Alberti's generative system, the height of the column is determined by the radius of its base, while in Gaudí's, the geometry of the church arches and vaults result from the behaviour of the weighted strings constrained under the force of gravity.

2.2   The Digital Condition

The introduction of computers in the 1950s opened new possibilities to explore calculated geometries in design. Inherent to its functioning, the computer is, first and foremost, a powerful calculation machine. That was the purpose that led a computer to be used for the first time in the field of architecture and construction, when Ove Arup engineers used a Ferranti Pegasus to calculate the structure of the Sydney Opera House in the early 1960s [8]. But, with the digital evolution, computers started to empower many other tasks, like representation and also manufacturing. In those early times, the aesthetic potential of geometric calculation through computation started to be examined by some artistic practices. Its distinctive features were pointed by Bruno Munari who, in the 1960s, explained that the ultimate aim of the Programmed Art was "the production not of a single definitive and subjective image, but of a multitude of images in continual variation" [9]. This formulation was enlightening, because it merged computing calculation and human creativity into a seamless artistic practice that defied established conventions. At the same time, other artists like Charles Csuri, in the same decade, started to examine the aesthetic and material potential of sculptures manufactured through digitally controlled machines as well [10].

In the last two decades, computational design and digital fabrication processes have become increasingly popular and decisive in architecture and construction. While conventional computer-aided design (CAD) processes tend to mimic the traditional manual production of drawings and models, computational design and digital fabrication processes allow designers to set parametric and algorithmic design strategies and, ultimately, to analyse and materialize them independently of their geometric complexity or variability. The irregular undulating shape of Zaha Hadid's Heydar Alyev Center in Baku, the non-standard structure of Herzog and de Meuron's Olympic Stadium in Beijing, the variable wooden components of Álvaro Siza, Souto de Moura and Cecil Balmond's Serpentine Pavilion in London, or the robotically assembled brick facade of Gramazio and Kohler's Gatenbein Winery in Flasch, are just a few examples, amongst many others today, which could not have been designed and built using traditional descriptive geometric processes. Despite their different scale and materiality, these four buildings share in common the generative exploration of geometry through calculation, both at the representation and fabrication levels. In sum, it seems plausible to affirm that through calculation and the unprecedented convergence of tools and techniques in a singular—digital—medium, computation is refactoring today the role and the impact of geometry in architecture.

3   Experiments in Architectural Education and Research

The context described in the previous section set the background of my research and educational activity over the last 15 years. During this time, I sought to examine the creative and material opportunities in architecture, emerging from exploring generative—calculated—geometry. To illustrate part of that experience, a selection of examples from education and research will be described. At the educational level, they were developed in several schools, like the Faculty of Architecture of the University of Porto (FAUP), the Department of Architecture of the Faculty of Sciences and Technology of the University of Coimbra (DArq/FCTUC), and the Institute for Advanced Architecture in Catalonia (IAAC). At the research level, they were developed in the scope of the Digital Fabrication Laboratory (DFL) research group, which I coordinate at the FAUP.

Given the different technological resources available in each of these academic environments, the examples presented in the current section employed different manual and digital strategies aiming to embrace calculated geometries in design. As Alberti, Guarino, Durand, Gaudí and Otto have shown before, there is no need for computers to generatively think and practice geometry. Nevertheless, given that human capabilities of calculation are limited, the use of computation dramatically expands the possibilities of such generative approach to geometry in architecture. These considerations will be hopefully clear with the experiments described next.

3.1 Education: "Parametric Architecture"—IAAC, 2008

Following the concept of parametric architecture presented for the first time by Luigi Moretti in 1966 [11], this assignment invited the students to think the design of buildings in terms of geometric parameters. By doing so with the help of computation, many different versions of a single design intention would be generated, evaluated and properly selected for the final solution(s). One of the works selected the Toló House designed by Álvaro Leite Siza in Vila Real and sought to describe the geometry of its volumes, their connections and relation with the topography, in terms of a system of variable and fixed geometric parameters (Fig. 1). Using the parametric design software TopSolid, the students were able to describe the original solution as well as many other versions adjusted to other programmatic specifications and/or topographic conditions. For each new set of inputs, the computer automatically calculated the geometry of a new design version. Unlike traditional explicit descriptive processes, this generative approach turns the design process flexible enough to consider (i.e. visualize, evaluate and select), at any time, different design possibilities.

3.2 Research: "Trefoil Structure"—DFL, 2013

This research work was inspired by a current trend in structural design seeking to achieve material optimization in structures. Taking advantage of parametric design and 3D printing technologies, the Ove Arup company has investigated the possibility of designing customized nodes in steel, employing the minimum material needed

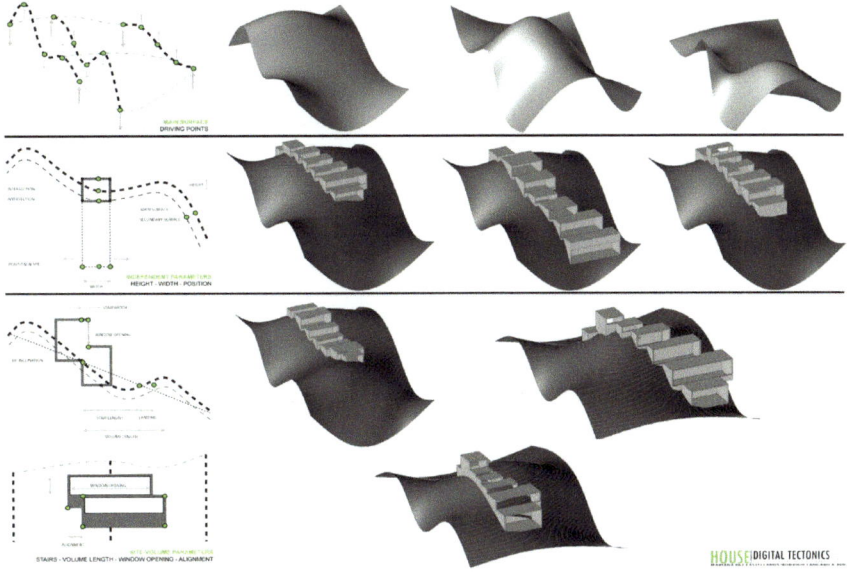

Fig. 1 Parametric design studies on the Toló House, showing the generation of different solutions according to the site topography (IAAC students: Mariana Paz and Rodrigo Lagarica)

for each location in the structure, and also, to fabricate them employing additive manufacturing processes [12]. This challenge for extreme design and material customization can only be attainable through the calculation of design geometries. With a similar goal in mind, the DFL developed a research on the three-dimensional geometry of a trefoil-knot. Using Rhinoceros™ software with the Grasshoper™, an initial surface of the Trefoil was converted into a grid of lines, which was then traduced into a set of tubular members connected with finger-joint nodes. Because the whole system is parametrically driven, many different structural densities can be generated, with the computer automatically calculating and designing the correspondent structural members and nodes (Fig. 2). In the instance selected for the research, the Trefoil was made out of 320 tubular members and 120 finger-joints, varying in size and orientation. While the production of the tubular members can take advantage of common digitally driven cutting processes, the volumetric joints can only be manufactured using 3D printing processes. A physical prototype of the structure was produced using plastic straws and PLA joints (Fig. 3). A single parametric model was the sole representation that generated, through calculation, all the geometries needed to design and build the Trefoil structure.

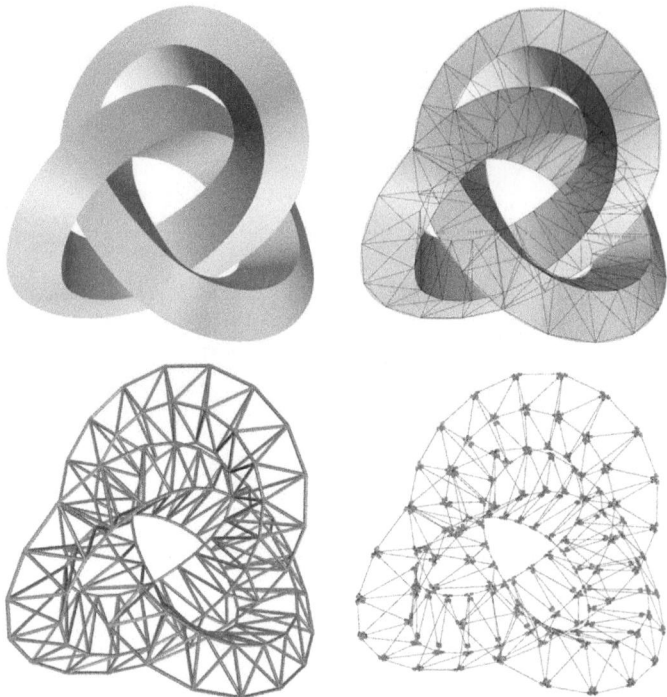

Fig. 2 Parametric development of the trefoil structure: the original surface and the generation of a structural grid (top), and the generation of the tubular members and joints (bottom)

3.3   Education: "Parametric Constructions"—FAUP, 2010–16

Taking place in the Constructive Geometry class, the investigation on parametric constructions unfolds in two assignments that explore the generation of parametric patterns in the 2D plane and in the 3D space. This work resonates to the "Parquet Deformation" studio taught by William Huff at Carnegie Mellon in 1966, when, without using computers, such adaptive design concepts were already thought of and exercised [13]. Now, using Rhinoceros and Grasshopper, students were challenged to develop parametric patterns capable of adaptation through progressive geometric differentiation of their module. With this experience, they had to reflect and imagine opportunities to integrate such strategies in architecture, for instance, in the design of facades or structures. In both assignments, the design explorations had to find a way to produce a physical model, prototype or installation (Figs. 4 and 5). While in the first editions of this investigation there was no digital fabrication equipment available, students followed manual cutting processes with the help of drawings generated by the computer. More recently, the exploration of digitally driven cutting processes made easier the materialization of such design explorations. The calculation of the

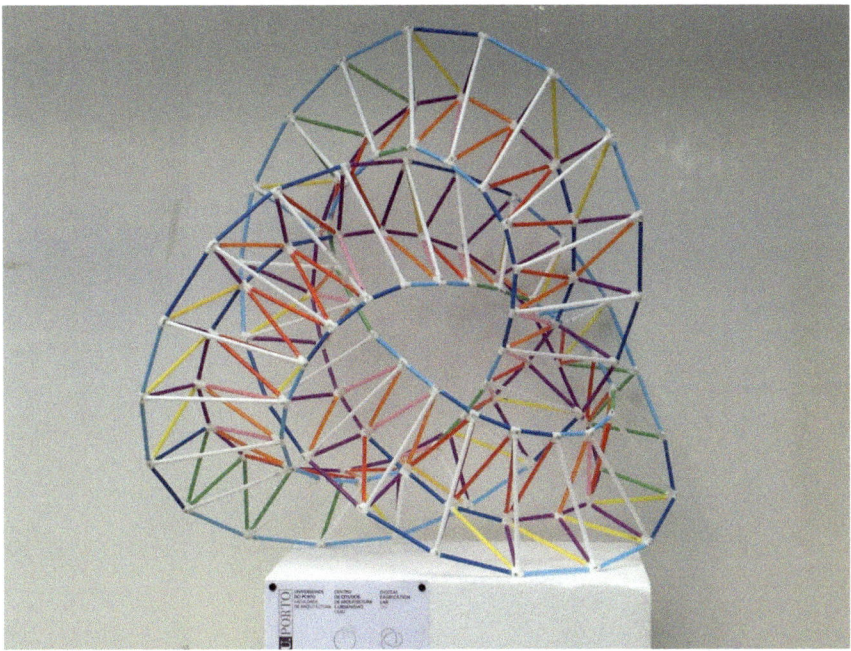

Fig. 3 The built model of the trefoil structure

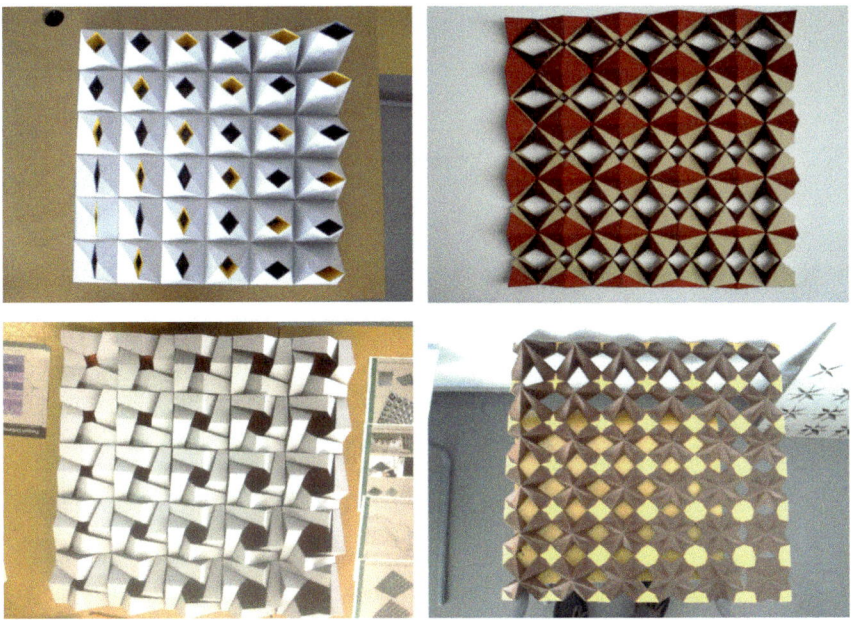

Fig. 4 Models of 3D parametric patterns generated from a 2D plane developed by the students

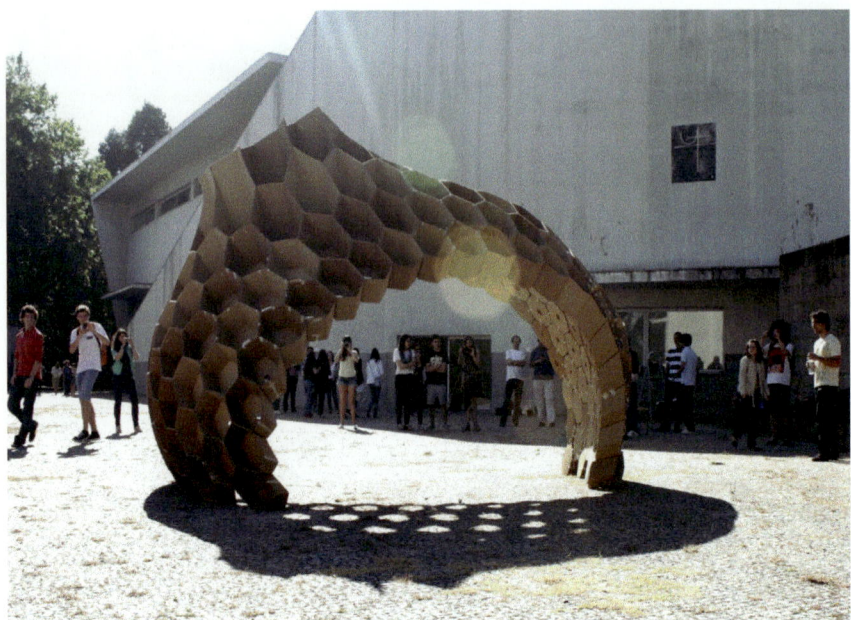

Fig. 5 Pavilion built at 1:1 scale at FAUP, based on a parametric pattern in the 3D space

geometries through computation is decisive not only for design, but also for the fabrication process.

3.4 Education + Research: "Robotic Assemblies"—FAUP, 2015–16

Besides investigating the use of computers in the design of a variable brickwork structure, this assignment aimed at using a robotic arm for its final material assembly, following the reference to the programmed walls developed by Gramazio and Kohler at the ETHZ [14]. Counting with 500 bricks at their disposal, the students of the Constructive Geometry class had to design a non-standard structure that could challenge the limits of manual brick layering [15]. To do so, it was not possible to imagine and model such structures brick by brick. Besides the difficulty in achieving structured and interesting results, it would be a nightmare to introduce changes along the design process. The right approach for such problem consists in defining parametric design strategies while the computer calculates the specific position of each brick in the structure. The spatial coordinates of each brick are then used to programme the sequence of pick-n-place movements of a robotic arm. In the class, the CG Column was selected among the different brick structures proposed by the students, to be robotically fabricated at the 1:1 scale (Fig. 6). The assembly process

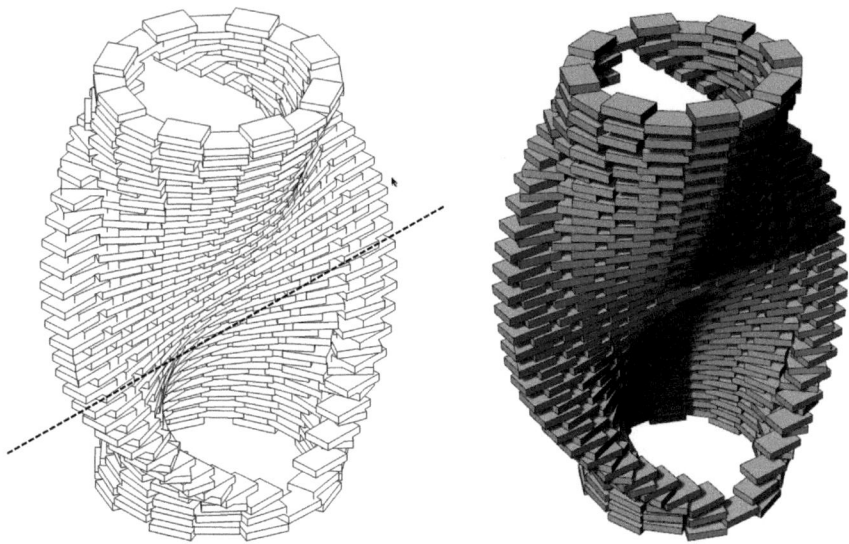

Fig. 6 The digital model of the CG column that was selected by the students to be robotically fabricated (students: Saule Grybenaite and Jorge Juan Pérez)

of the subtle variation of its geometry resulted from a calculation procedure that bridged its design to its robotic fabrication (Fig. 7). The built result would almost be impossible to achieve, with such precision and on time, if the process had to rely in a set of descriptive metric drawings.

A similar experiment was conducted at the DFL. Departing from brickwork details presented in buildings designed by the Portuguese architect Raúl Hestnes Ferreira, the DFL team explored the possibility of expanding the design intention of the architect using parametric design and digital fabrication processes [16]. After conducting a series of studies on column designs, a twisted solution was selected by Raúl Hestnes Ferreira to be robotically fabricated at the 1:1 scale. With such digital process, it took just 50 min to assemble and glue the complete Hestnes column (Fig. 8). Independently of its geometric complexity, any other column design with the same number of bricks would have taken the same time to be assembled. The same could not be said, though, about manual assembly processes based on reading descriptive drawings.

3.5 Education: "The Curved Building"—FAUP, 2010–16/DArq-FCTUC, 2010–12

The final experiment differs from the previous ones, as it does not rely on parametric design processes. Motivated by manual paper bending techniques like those explored by Frank Gehry to conceive his designs, this assignment challenged the students to

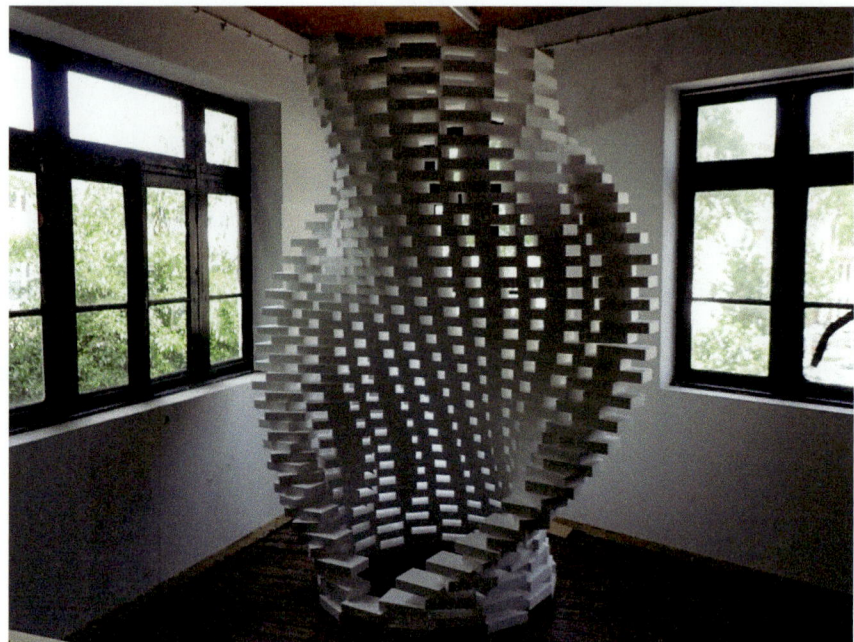

Fig. 7 Robotically assembled CG column

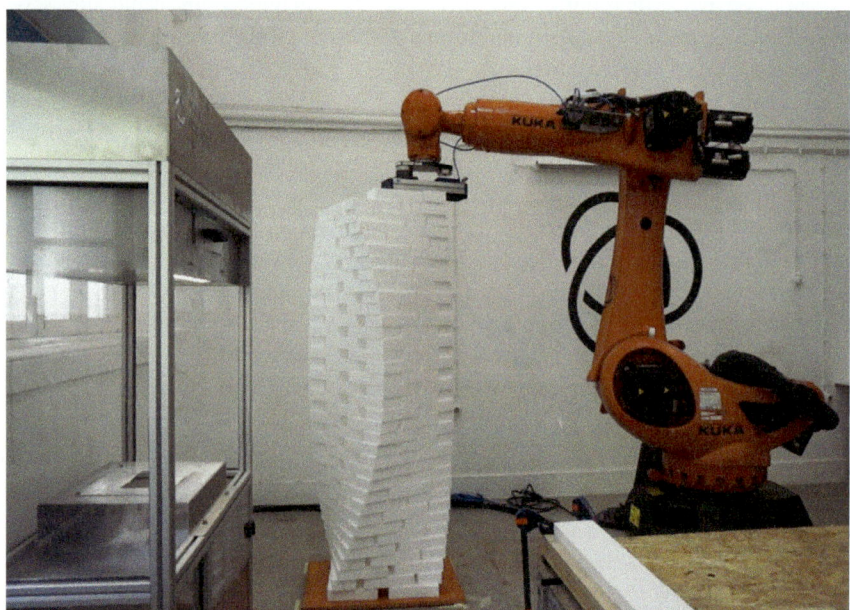

Fig. 8 Hestnes column assembled by the robot at the DFL

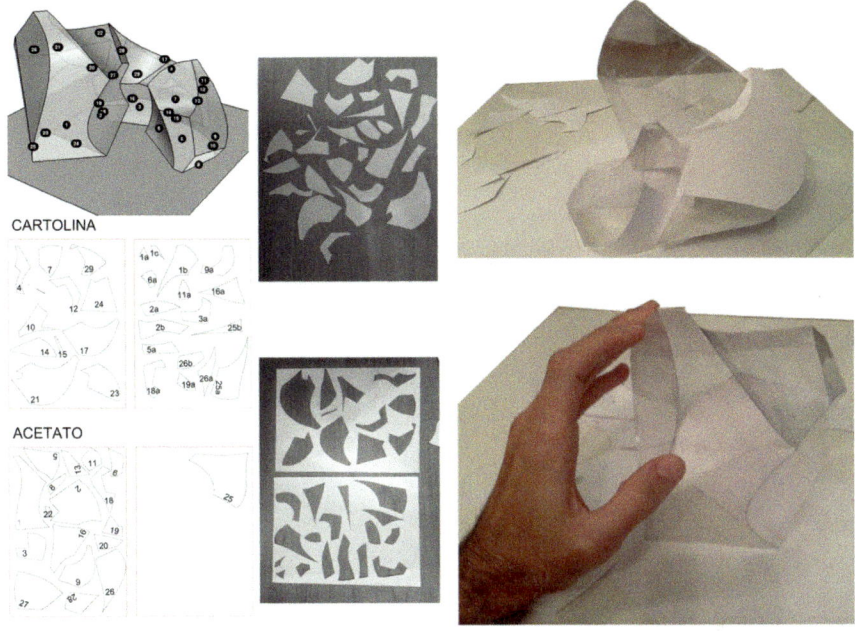

Fig. 9 Fabrication of the curved models with the help of the computer for flattening and cutting the pieces (student: Caio Cavalcanti)

conceive and construct curved geometries based on ruled surfaces with the help of computers [17]. The work unfolds in the computer by simulating that material-based design process to conceive a volumetric composition of curved *building* shapes. And, for its materialization, students have to find the flattened version of all the curved surface geometries for cutting. Since it would be impossible to obtain those flattened versions through geometric drawing constructions, they have to rely on a computer algorithm for their calculation. Unlike Frank Gehry, who is conceiving through direct manipulation of the final physical parts, the students are manipulating representations that demand finding other procedures for their materialization. In the end, none of the solutions achieved with this assignment could be modelled and fabricated through conventional descriptive projection drawings (Figs. 9 and 10).

4 Conclusion

Considering the initial distinction between *descriptive* and *generative* modes of embracing geometry in design, the experiments in architectural education and research described above clearly fall in the latter category. Due to the complexity of the different geometric challenges, the use of digital parametric and fabrication processes was fundamental to represent and materialize such design intentions. By

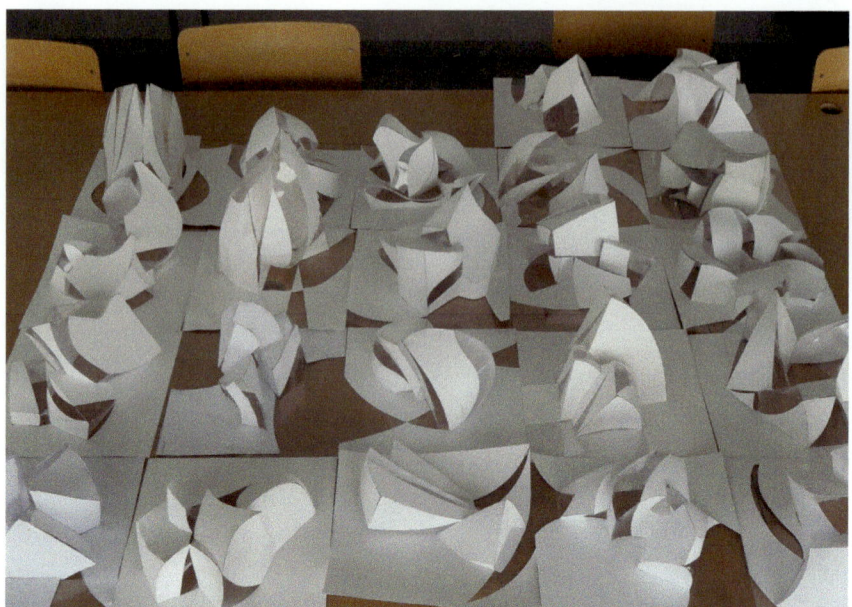

Fig. 10 Curved buildings digitally conceived by the students

tying the capabilities of representation and calculation, they reinforce and extend the role of geometry in architecture, becoming decisive skills for any architect and designer today. I hope the experiments described may contribute to enlighten ways of integrating them in the curriculum of architecture.

Acknowledgements I would like to express my gratitude to the schools that provided the academic and research conditions to develop the works presented in this text, namely, FAUP, DArq/FCTUC and IAAC. I would also like to thank my colleagues with whom I had the pleasure to share some of these academic experiments, namely to Marta Malé-Alemany in the direction of the Parametric Design studio at IAAC, to João Pedro Xavier in the Constructive Geometry class at FAUP, and also to the DFL team members Cristina Gassó, Manuel Oliveira, Pedro Martins, Pedro Varela and Rui Oliveira, who participated more actively in the research works presented in this article.

References

1. Mitchell, W.J.: Roll over Euclid: how Frank Gehry designs and builds. In: Fiona Ragheb, J. (ed.) Frank Gehry Architect. Solomon R. Guggenheim Foundation, New York (2001)
2. Dürer, A.: Vier bücher von menschlicher proportion (approx. 1528)
3. Alberti, L.B.: De Re Aedificatoria (approx. 1452)
4. Roero, S.: Guarino Guarini and universal mathematics. Nexus Netw. J. **11**(3), 415–439 (2009)
5. Durand, J.N.L.: Précis des leçons d'architecture données à l'École polytechnique (approx. 1802)

6. Huerta, S.: Structural design in the work of Gaudí. Architectural Sci. Rev. **49**(4), 324–339 (2006)
7. Otto, F.: Complete Works: Lightweight Construction—Natural Design. Birkhauser, Basel (2005)
8. Watson, A.: Building a Masterpiece: The Sydney Opera House. Lund Humphries Publishers, UK (2006)
9. Munari, B.: Arte Programmata. In: Armstrong, H. (ed.) Digital Design Theory, p. 29. Princeton Architectural Press, New York (2016)
10. The Charles A. Csuri Project. Visited in October 2017 at: http://www.csuriproject.osu.edu
11. Bucci, F.: Luigi Moretti. Princeton Architectural Press, New York (2002)
12. Ove Arup: Design method for critical structural steel elements. Visited in October 2017 at: http://www.arup.com/projects/additive-manufacturing
13. Huff, W.S.: What is basic design? In: Crowell, R.A. (ed.) Intersight One. State University of New York, Buffalo (1990)
14. Gramazio, F., Kohler, M.: Digital Materiality. Lars Muller, Zurich (2008)
15. Sousa, J.P., Xavier, J.P.: Fabricação robótica de estruturas em tijolo. Uma experiência no ensino da arquitectura. In: Proceedings of the SIGRADI 2015, pp. 143–147. Florianópolis, Blucher Design (2015)
16. Oliveira, R., Sousa, J.P.: Building traditions with digital research. Reviewing the brick architecture of Raúl Hestnes Ferreira through robotic fabrication, pp. 123–131. University of Oulu (2016)
17. Shelden, D.: Digital surface representation and the constructability of Gehry's architecture. Ph.D. thesis in Architecture. Massachusetts Institute of Technology, Department of Architecture (2002)

José Pedro Sousa is an Assistant Professor at FAUP (University of Porto), where he founded and directs the DFL, Digital Fabrication Lab (CEAU/FAUP). He has a Ph.D. degree in Architecture from IST (University of Lisbon), a Master in Genetic Architectures from ESARQUIC (International University of Catalonia) and a Licenciatura in Architecture from FAUP. He was also a Special student on Design and Computation at MIT (Massachusetts Institute of Technology, USA) and a Visiting Scholar in Architecture at the UPenn (University of Pennsylvania, USA). With an interest in exploring new conceptual and material opportunities emerging from the use of computational design and fabrication technologies, he has developed a recognized professional activity merging the realms of teaching, research and design practice since 2003. He was awarded with the 2005 FEIDAD Outstanding Award (1st) and the 2009 Young Research Award of the Technical University of Lisbon, among other distinctions. José Pedro Sousa was one of the keynote speakers of the international conference Geometrias'17.

How to Construct the Red Sea?

Monika Sroka-Bizoń

Abstract The possibility of realization of freeform shapes in architecture is, in some way, a symbol of contemporary architecture. Wavy roofs and walls of buildings, curvilinear shapes of objects designed and executed through CAD techniques, are now likely to be accomplished easier, quicker and cheaper than in the past. So, we ask ourselves, is it possible to build the *Red Sea?* The Museum of the History of Polish Jews built in 2013 in Warsaw, stands as one of the newest Polish accomplishments, dealing with curved surfaces in architecture. The main part of the museum's building is a cuboid interrupted by a curvilinear rupture. Architects describe this curvilinear part in the project as the *Yum Suf, or Red Sea.* Details on its concretization are of great interest for the author of this paper.

Keywords Geometry · Architectural geometry · Design · Architectural design · Freeform surfaces

1 Introduction

In the spring of 2013, the four-year construction of this important public utility building was completed and the Museum of the History of Polish Jews opened to the public. The museum is situated in the Warsaw's district Muranów. The idea for creating this *Museum of Life* in the capital of Poland, as it has been named from the beginning, was initiated in 1993 by the Association of the Jewish Historical Institute of Poland [1]. Before the Second World War, Poland was the centre of the Jewish Diaspora and home for the largest Jewish community in the world. During the inter-war period, Poland was inhabited by 3–3.5 million Jews. In Warsaw alone, Polish Jews comprised almost one-third of the city's population. The Second World War brought tragedy upon the Jews of Poland. A huge part of Polish Jews perished during the Holocaust, and part of those who survived emigrated from Poland after the War.

M. Sroka-Bizoń (✉)
Department of Building Engineering and Building Physics, Faculty of Civil Engineering, Silesian University of Technology, Gliwice, Poland
e-mail: monika.sroka-bizon@polsl.pl

© Springer Nature Switzerland AG 2020
V. Viana et al. (eds.), *Thinking, Drawing, Modelling*,
Springer Proceedings in Mathematics & Statistics 326,
https://doi.org/10.1007/978-3-030-46804-0_11

But despite all odds, the Polish Jewish diaspora survived, and above all, the memory of the Polish Jew's rich culture prevailed. The thousand-year history of the Polish Jews and its impact in Poland in the past and nowadays was a major impulse to create the Museum of the History of Polish Jews. In 1995, the Warsaw City Council allocated the land for this purpose, and the area designed for the museum was located in a small park in Muranów, the district of Warsaw, where, during Second World War, the Warsaw Ghetto was situated. The plot intended for the construction, a square between Anielewicza, Zamenhofa, Lewartowskiego and Karmelicka streets, was an important part of the Ghetto. In 1948, the Ghetto Heroes Memorial was positioned in this area [2].

In January 2005, the city of Warsaw, the Polish Ministry of Culture and National Heritage, and the Association of the Jewish Historical Institute of Poland signed an agreement establishing a joint cultural institution—the Museum of the History of Polish Jews. It was then determined that the choice of the museum's design would be decided through an architectural competition. According to the terms and conditions of the competition, the new building should be functional and modern and include a recognizable characteristic form, in order to become one of the symbols of contemporary Warsaw; also, the new facilities should not interfere too much with the existing space located in the open green area shared with the residents.

The International Architectural Competition for the design of the museum was launched in February, 2005. The competition attracted great interest from the architectural designer's milieu. More than 200 architects, from 36 countries, registered for the first part of the competition. For the second part of the competition, the jury chairman, Bohdan Paczowski, invited eleven architectural teams led by the following architects: Andrzej Bulanda (Bulanda and Mucha Architects, Poland), David Chipperfield (David Chipperfield Architects, Great Britain), Marek Dunikowski (DDJM Architectural Studio, Poland), Peter Eisenman (Peter Eisenman Architects, USA), Zwi Hecker (Zvi Hecker Architect, Israel/Germany), Kengo Kuma (Kengo Kuma & Associates, Japan), Daniel Libeskind (Daniel Libeskind Studio, USA), Rainer Mahlamäki (Lahdelma & Mahlamäki Architects, Finland) Josep Luis Mateo (MAP Architects, Spain), Jesus Hernandez Mayor (Casanova + Hernandez Architects, Spain) and Gesine Weinmiller (Weinmiller Architekten, Germany) [3].

On June 30, 2005, the results of the competition were published. It was won by the Finnish practice of Lahdelma & Mahlamaki Architects that consists of two Finnish architects—Mahlamäki and Lahdelma [4]. This was the first victory of the Finish Studio in an international competition. The architectural project of the museum was developed in international cooperation with the Polish architectural studio Kuryłowicz & Associates, led by the architects Stefan Kuryłowicz, Ewa Kuryłowicz, Paweł Grodzicki, Marcin Ferenc, Piotr Kuczyński, Tomasz Kopeć, Michał Gratkowski and Piotr Kudelski. ARBO Projekt with Arkadiusz Łoziński, Robert Fabisiak and Piotr Ziółkowski developed the construction part of the design. The general contractor of the building was the company Polimex-Mostostal. The company TORKRET made curvilinear walls using spray-concrete technology. In the summer of 2009, the construction of the building was initiated, and it lasted four years [5].

2 Concept of the Museum

The building of the museum was located into the green open area opposite to an important war monument, the Ghetto Heroes Memorial. The terms of the competition stated that the museum's building could not dominate this monument. The building area of the museum could not occupy more than one-third area of the square, and the buildings' height should not exceed the surroundings blocks of flats. Taking these constraints in consideration, the museum is situated in the central part of a trapezoid building plot with an area 4.40 ha. The building of the museum is a square based cuboid, with 67 m of edge length. The height of the aboveground part of the building is 21 m. The museum has two underground levels, four of them above ground, and 16,000 m^2 of useable space, of which one quarter is dedicated to the main exhibition. The outer shell of the museum is a structure composed of vertical panels of perforated copper sheet and point-fixed tempered glass. Glass panels of the building's facade are 0.44 × 1.7 m wide, and copper panels are twice narrower. These panels are suspended on a lightweight steel structure fixed to the reinforced concrete walls of the building, creating alternated glass and copper straps aligned to each other at right angles. Thanks to this, the building facades gain depth and change their character, depending on the location from which they are observed. The facades are visible as bright glossy surfaces from one point; and, from another point, as if they were almost completely transparent. Only individual windows of selected rooms cross the shell [6].

The parallelepipedal object, due to its scale and simplicity, is very well inscribed in the contemporary urban context of the location and the surrounding housing of the 1960s. The blocks of flats that surround the museum are simple cuboidal objects, and likewise, the shape of the museum is cuboidal. Nevertheless, the chairman of the jury, Bohdan Paczowski, named the concept of the museum proposed by the Finish architects, as *a cookie in a box* [5]. The cuboidal form of the building is the *box*, for sure, but where can the *cookie* be found? If one looks at the project's situational plan, on the rectangular outline of the building, some corrugated lines are visible, and these have been described by the authors of the design as *Yum Suf*. *Yum Suf*, in Hebrew, means the *Red Sea*, but in the biblical sense, meaning the parting of the *Red Sea* before the wandering of Jewish people. A closer look at the conceptual sketches and the physical model of the object allows us to understand both the authors' description and the opinion of the chairman of the jury. The curved walls of the inner passage interrupt the cuboidal object of the museum, so the *cookie* is to be found *inside the box* (Figs. 1 and 2).

The curvilinear rupture of the building, according to the architect's description in the design, symbolizes the biblical red sea—*Yum Suf*—which is parting, for the safe escape of Jews from Egypt. The expressive break in the cuboid opens the building from the side of the Ghetto Heroes Memorial, which is situated on the east side of the plot, to the green open area, in the west part of the plot.

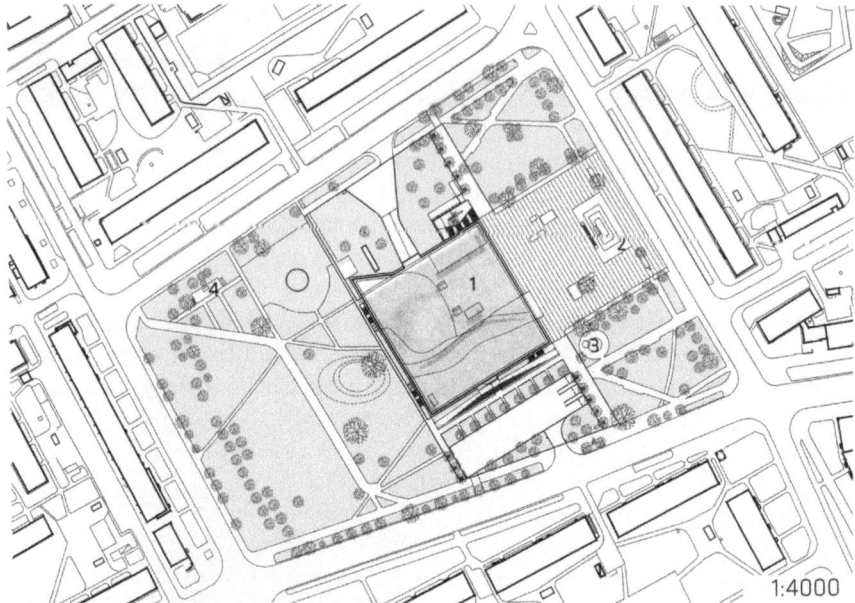

Fig. 1 Location plan of the Museum of History of Polish Jews, arch. Lahdelma & Mahlamaki (Finland), Kuryłowicz & Associates (Poland). In the rectangular outline of the building, corrugated lines, described by the authors of the design as *Yum Suf*, are visible. Image elaborated by the author, based on [5]

3 Concept of the Curvilinear Wall

Curvilinear walls in the inner passage rise from the underground exhibition's space of the museum to the roof. They divide the object almost on all of its height. The architectural and construction design of these curvilinear walls was developed in four stages:

- freehand sketch prepared by the architect Rainer Mahlamäki,
- scan of the architect's freehand drawing,
- computer model worked out in AutoCAD™
- computer model worked out in Rhinoceros™

The construction design of the walls was a great challenge for the engineers from ARBO's design studio, particularly, the walls, that were conceived as load-bearing walls, on which the floors of the building are supported and, in some places, the roof of the building as well. In this case, it was necessary to prepare the construction design of two curvilinear walls with more than 20.0 m of height and free-shaped curvatures. In the initial design, assumption walls were to be constructed as monolithic concrete walls, and these can be formed in milled formworks. Their design and execution do not cause major difficulties, if their shape is constructed from a ruled surface, allowing thus for a relatively simple ruled surface to be constructed as a reinforced

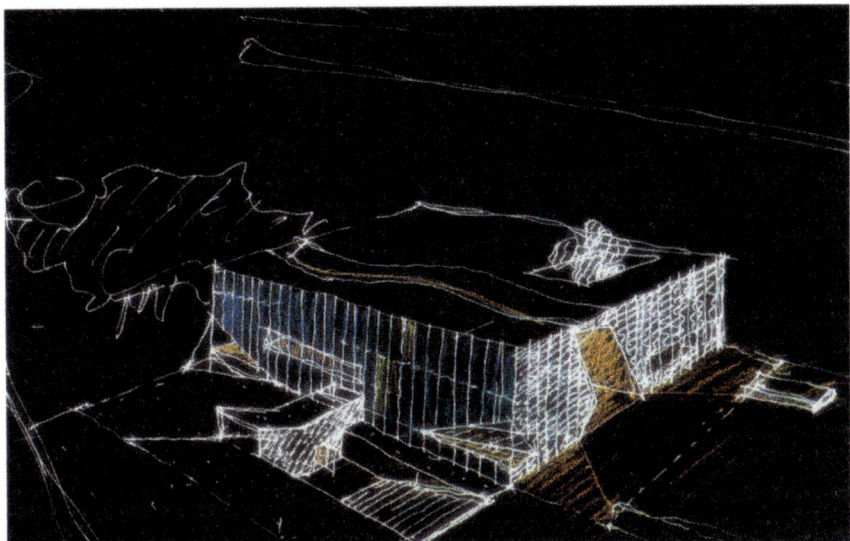

Fig. 2 Conceptual sketch of the Museum of History of Polish Jews, arch. Lahdelma & Mahlamaki (Finland), Kuryłowicz & Associates (Poland). In the rectangular outline of the building, corrugated lines, described by the authors of the design as *Yum Suf*, are visible. Image elaborated by the author, based on [5]

concrete shell. Consequently, the main reinforcement rods for the structure are the surfaces' generating lines. A different design issue, however, was the project of the formwork for the reinforced concrete shell in such cases. In this particular case, the free-shaped curvature of the walls made it impossible to design a repeatable milled formwork for the concrete structures, because that would significantly increase the cost of construction. Therefore, the construction of the wall was changed: instead of a reinforced concrete structure, a steel structure with pipes was used, and steel pipes form the skeleton construction of curved walls.

One problem of the design has thus been solved—the construction of load-bearing walls of complicated geometrical structure, through a more economic version, but another occurred: how would the surface of the curvilinear walls be modelled? This problem was solved by designing mesh elements for the surface. Engineers and architects decided to design the curved surface of the walls as an approximation of the mesh, using quadrilateral planar faces. The digital walls model was prepared in the software Rhinoceros in order to outline the geometrical shape of each panel and the geodetic coordinates of each corner of the panels. In the primary version of this part of the constructional design, modelling the surface of the walls was planned through resin cement panels cast in situ, that were then incorporated into the steel curvilinear wall structure with a system of holding elements. However, because of the problems in obtaining panels made of resin cement with adequate quality, and problems with suitable finished connections for the wall panels, the technology for the construction of a curvilinear wall was rejected. With a firm knowledge on the possibilities of

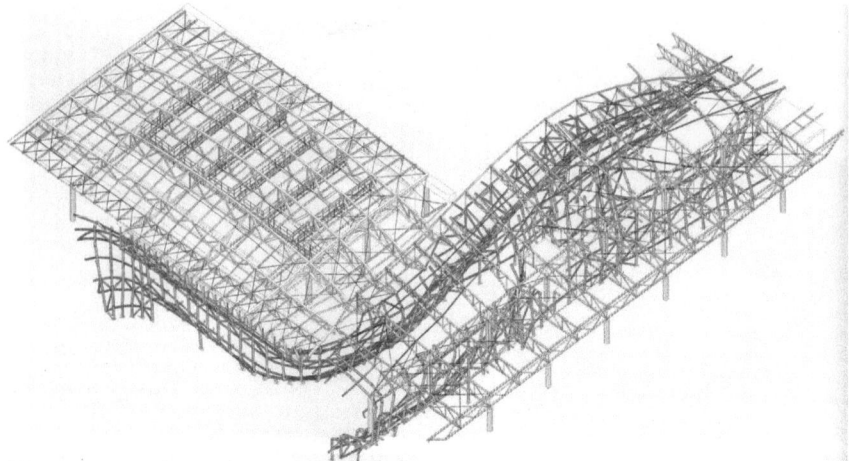

Fig. 3 3D model of the steel structure of the curvilinear wall. Design of construction: ARBO Projekt: Arkadiusz Łoziński, Rober Fabisiak, Piotr Ziólkowski; Museum of History of Polish Jews, arch. Lahdelma & Mahlamaki (Finland), Kuryłowicz & Associates (Poland). Image elaborated by the author, based on [5]

producing curvilinear surfaces with shotcrete technology, the company TORKRET prepared 3D models of curvilinear wall as an illustration of the solution that could be used for the intended design. According to this technology, wooden panels made of OSB plate were used as a first layer that was then fixed to the steel structure. The following layer was reinforced fabric, and the structure's last layer was shotcrete, that is, concrete applied through spraying. The final drawing of the meshes of the wall's surface, meant to be visible, was obtained with the help of real and apparent expansion joints. This way, in the final solution adopted for the problem of modelling the curved wall surfaces, the wall resin cement panels made in situ were replaced by wooden panels made of OSB panels, that had been covered with a layer of shotcrete [7] (Figs. 3, 4 and 5).

4 Conclusions

1. The accomplishment of the Museum of History of Polish Jews demonstrates that the complete design and construction process of a building, as complex as this, involves many aspects, that go from the architects' idea of the design, to form-finding and geometrical structure, statics, materials, feasible segmentation into panels and the costs of the construction.
2. The construction of the geometrical model of the curvilinear walls was based on the free-shaped curvature of the walls. Shaping curves in CAD programs may depart from their parametric representation, being the most common types:

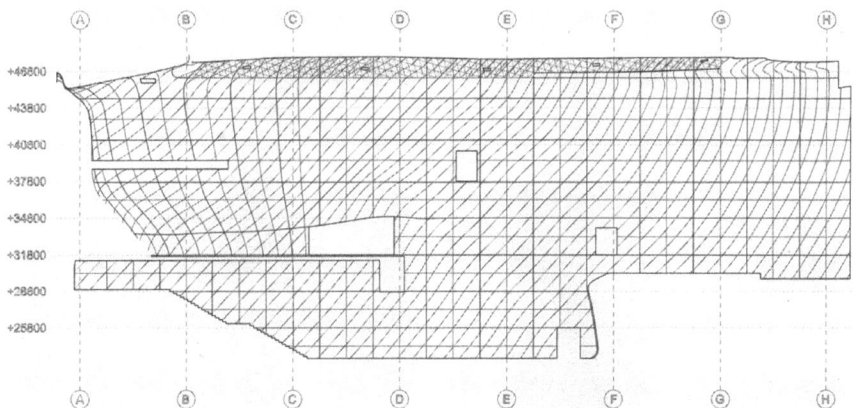

Fig. 4 Mesh of the curvilinear wall. Museum of History of Polish Jews, arch. Lahdelma & Mahlamaki (Finland), Kuryłowicz & Associates (Poland). Image elaborated by the author, based on [5]

Bezier's curves, B-splines and NURBS curves. Computer models of surfaces based on these curves can, respectively, be: Bezier surfaces, B-spline surfaces and NURBS surfaces. The materialization of such surfaces is based on discrete surface theory, and the most basic, conventional and structurally stable way of representing their smooth shape in a discrete way, is via the use of triangular meshes. Alternative ways of representing the smooth shape of a surface might be through quadrilateral meshes with planar faces, as exemplified in the analysed construction [8].

3. The construction of geometric computer models of freeform surfaces through contemporary CAD programs is currently not a problem for designers; the bigger problem is to find a good design execution solution for the project. Problems found in the execution of the project analysed were: a feasible segmentation of the surface into panels and the materialization of these segmentations; the search for the best materials for the accomplishment of the main idea for the design and solving these predicaments in an economical way, which turned out to be far more difficult than the computer modelling of the freeform surfaces.

4. Discrete models of freeform surfaces may be an important part of the structure in architecture, not merely a constructional part of the project. The discrete model of the surface, the transparency of the panels and the shape of the approximation of the surface, can be important parts of the solution, and so it was in the object analysed [8] and [9].

5. The technological and economical aspects adopted in the design solution and the discrete models of the curvilinear surfaces were obtained with layers of shotcrete.

6. Further analysis of the designing and realization processes of the curvilinear walls of the Museum of History of Polish Jews allowed us to conclude that, regardless of highly developed computer-aided design techniques, basic issues

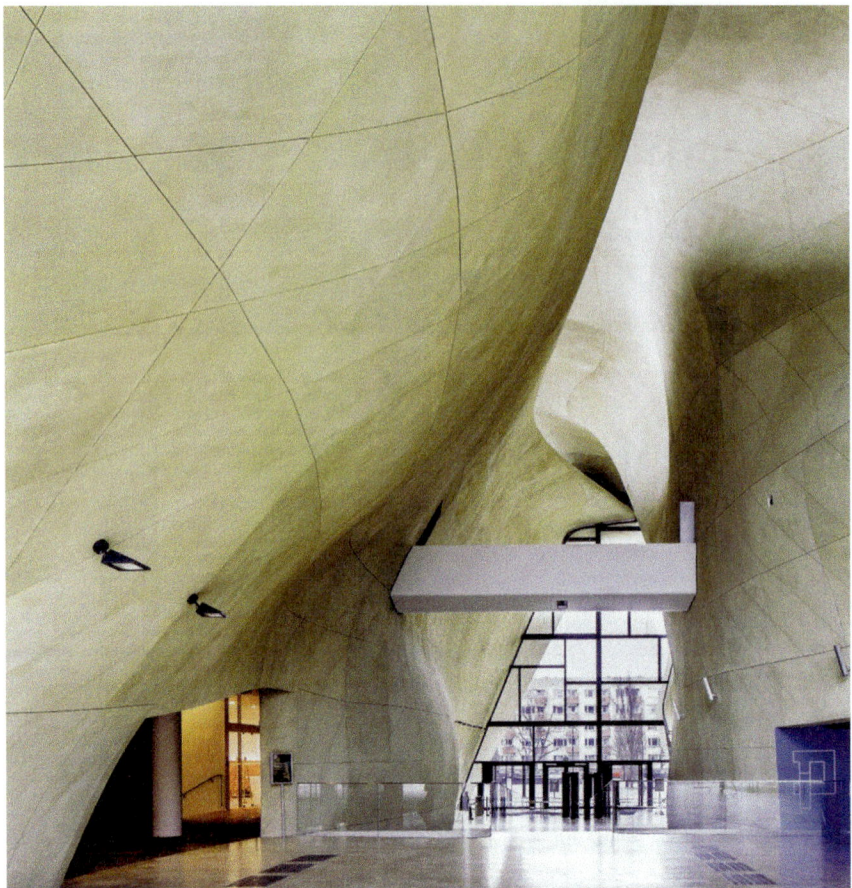

Fig. 5 Inner passage, view towards to the main entrance. On the wall, the drawing of the mesh of the curvilinear walls is visible. Museum of History of Polish Jews, arch. Lahdelma & Mahlamaki (Finland), Kuryłowicz & Associates (Poland). Image elaborated by the author, based on [5]

in the design processes turn out to a problem that can be formulated as a short question—how to turn it?

References

1. Podgórska, J.: Muzeum życia. O tym jak powstało Muzeum Historii Żydów Polskich, jak zostało zorganizowane I jakie niesie przesłanie, opowiada Marian Turski. Polityka **43**, 2981 (2014)
2. Stiasny, G.: Getto, warszawska architektura pamięci. Architektura Murator **6**, 26–36 (2013)
3. Kiciński, A.: Konkurs na projekt museum historii Żydów Polskich w Warszawie - idea, twórcy, zadanie, wynik. Muzealnictwo tom 2005, No. 46 (2005)

4. Retrieved in June, 2017 from http://www.sarp.org.pl/pokaz/ilmari_lahdelma-rainer_ mahlamäki_z_finlandii-zwyciezcami_konkursu__na_muzeum_historii_zydow_pols,305
5. Cudak, A., Ferenc, M., Łoziński, A., Mahlamaki, R., Mycielski, K., Stiasny, G., Żylski, T.: Muzeum Żydów Polskich. Architektura Murator **6**, 38–60 (2013)
6. Retrieved in June, 2017 from http://architektura.nimoz.pl/2013/03/09/konstrukcja-muzeum-historii-zydow-polskich
7. Czajka, W.: The museum of the history of Polish Jews. Shotcrete Magazine, 2/26/2013
8. Pottmann, H., Asperl, A., Hofer, M., Kilian, A.: Architectural Geometry, pp. 669–709. Bentley Institute Press, USA (2007)
9. Fabiańska, P., Ferenc, M.: Rozstąpione Morze. Muzeum Historii Żydów Polskich w Warszawie. Świat Architektury **6**(36) (2013)

Monika Sroka-Bizon is an architect and a Tutor in the Faculty of Civil Engineering at The Silesian University of Technology in Gliwice, Poland. Her research interests are Architecture, Geometry and Graphics and their applications in Architecture.

How to Improve the Education of Engineers—Visualization of String Construction Bridges

Jolanta Tofil[ORCID]

Abstract In today's increasingly developed and built-up environment, Civil Engineering students face the problem of how to connect the technical intricacies of building with important aspects of design, in terms of form and aesthetics. The building structure is one of the main elements in the overall design process that has to be considered by architecture practitioners, as well as engineering practitioners. Unfortunately, in the majority of cases, structural design is carried out solely by engineers, and as such, they have an important role in determining which appropriate systems comply best with architectural design. It is the author's belief that the graduating students of Civil Engineering Faculty should be able to apply their knowledge of each technical system in the context of any architectural design project. Architects and engineers rely heavily on the use of computer tools and software in their work, especially computer-aided design (CAD) and visualization programs. In this study, we present the result of a connection between researches on the structure of string construction objects and education of engineers.

Keywords Spatial structures · String construction bridges · Visualization · Engineering education

1 Introduction

Due to the current rapid development of global economy and the increasing number of road investments in Poland and the world, engineering constructions as bridges are fairly common sights, and almost every day we walk or drive over a bridge. Their role is undisputable, as they serve the dual purpose of facilitating journeys and enabling fast-trade exchanges. Thanks to nowadays technological developments, contemporary engineers delight us with unusually slender string constructions created with modern materials in order to overcome difficult and several-kilometer-long obstacles. Bridges constitute a vital part of our building space, and should therefore be

J. Tofil (✉)
Faculty of Civil Engineering, Silesian University of Technology, Gliwice, Poland
e-mail: jolanta.tofil@polsl.pl

© Springer Nature Switzerland AG 2020
V. Viana et al. (eds.), *Thinking, Drawing, Modelling*,
Springer Proceedings in Mathematics & Statistics 326,
https://doi.org/10.1007/978-3-030-46804-0_12

155

the subject of more investigations, in order to determine which designs are more appropriate, in terms of construction, as well as aesthetics. These arguments lead us to the conclusion that this type of construction should be chosen as one of the main tasks to be proposed to the future civil engineering that we teach.

The aforementioned economic development has facilitated a dynamic development of information technologies. Innovative computer software generates, in turn, rapid developments in every field of life. Unsurprisingly, geometric spatial modeling tasks must be included in the scope of every engineer and architect interests. Every day, we observe an increasing implementation of cinematographic virtual images, some of which are created for computer games. To conduct the classes of "Visualization of Buildings" with the eighth semester students of the Faculty of Civil Engineering, we have chosen the software 3ds Max^{TM}. In this program, a number of animations and videos have been created, as, for instance, the Polish "Cathedral" of Tomas Baginski. However, this software is mostly used to visualize architectural objects. Also, the problems of simulation and animation which can be achieved through this tool are of great interest.

2 Creation

3ds Max is a very extensive program, with many tools and specific techniques to create the individual elements of a given structure. Each element can be generated in several ways depending on the goal intended for the model under analysis. The general opinion about the software 3ds Max is that it is very complex and difficult to operate with, being this the reason why some beginners in the field of visualization choose simpler programs like SketchUp^{TM}. In our opinion, this is not true, since to perform basic model in these two programs takes almost the same amount of time and skills. Furthermore, the differences which distinguish the models accomplished in each of the programs can be seen when we want to get better visualization results. Significant impact on the quality of the model and subsequent reception of visualization are created through accuracy, the amount of detail, and texture's quality. These elements make the models created in simpler programs reach their limits and not being able to meet the expectations and effects of a more accurate design. Despite the possibility to import such a model to a more advanced software, the models turn out to be completely unsuitable for further processing and modification, a problem that is often associated with the creation of the model from scratch.

3 Education

3ds Max is so wide-ranging that it is impossible to show all of its features during the 30 h of laboratory classes. Therefore, a selection of individual issues was made, in order to grant priority for the tools most commonly used in architectural modeling, particularly, the most useful and versatile. As an introduction, students were trained in the acquisition of 3D modeling skills using such techniques as extrusion, polygon modeling, and Boolean operations. In the case of bridges, which are located over various obstacles, much attention was devoted to terrain modeling (Fig. 1a, b).

For the next stage, students dealt with the design preparation for visualization and animation of objects and views. The knowledge gained in the process was supplemented by the issue of creating and texturing, as well as introducing lighting features in the scene. For the day mode, a daylight system was used, while, to simulate the night, lighting photometric light was used, configured according to different profiles (Fig. 2a, b).

It was also significant for students to learn how to create and customize cameras for future rendering and animation. It is worth mentioning that Mental Ray engine was used for the renderings, and therefore, the more suitable materials for this were chosen.

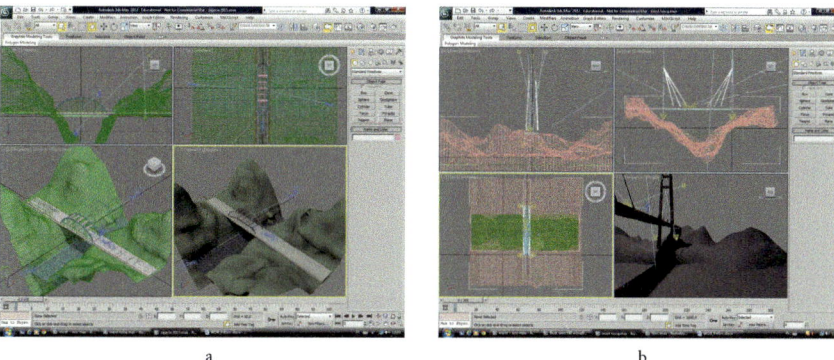

a   b

Fig. 1 Terrain modeling made by students during visualization of buildings classes

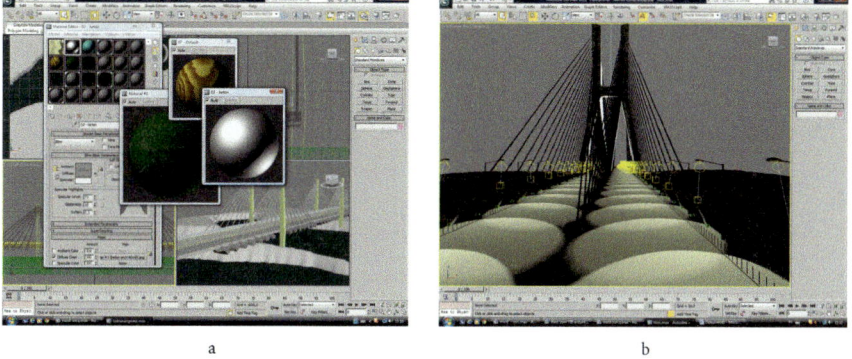

a   b

Fig. 2 Texturing and lighting of the scene made by students during visualization of buildings classes

Students had an extensive lecture on the principles of the operation of string arrangements in bridges and on structures, such as rigging systems, pylons, and supports, as well as common building materials before laboratory classes. Also, numerous examples of these types of objects already built, along with architectural and construction drawings and visualizations created at the design stage, were presented. During the presentation, the aims and formal requirements of getting credits from these classes were gradually discussed. After the lecture, the academics responded to various questions about technical and engineering issues, software and hardware, as well as organizational matters. The students received well the idea of selecting a designing example for the project, which was subsequently worked during laboratory classes, and even suggested organizing a competition for the best visualization at the end of the semester.

4   Representation

Any architectural modeling of a building structure requires presenting it in a virtual space, which is a representation of the real environment where it is to be accomplished (Fig. 3a, b).

At the stage of computer modeling, it can be already predicted how the building will fit with the surrounding natural or urban landscape. This allows multiple possibilities of changing decisions for the applied design structure or the shape of the geometric form of the object's individual elements.

To assess the students' accuracy in applying the structure in a virtual space, they were asked to perform three takes within the developed renderings. The first rendering, in bird's-eye view, should involve the whole design. The second, from the user's position, of a person who passes through the bridge. The third one, from an observer's position located under the bridge level, as if seen from a boat. All of them should be rendered on day-and-night mode (Fig. 4a–f).

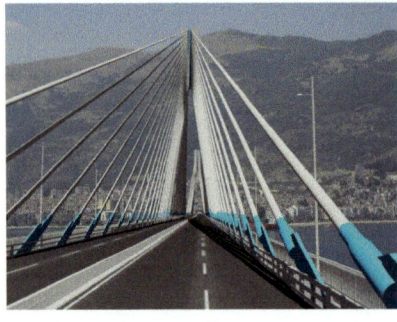
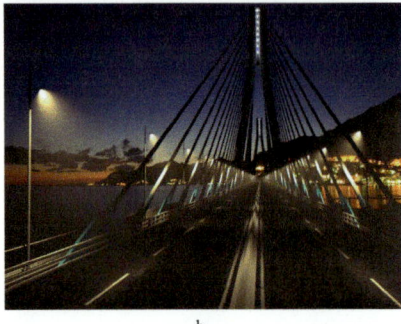

a   b

Fig. 3 Setting of the bridge in the environment where it is meant to be accomplished, made by students during visualization of buildings classes

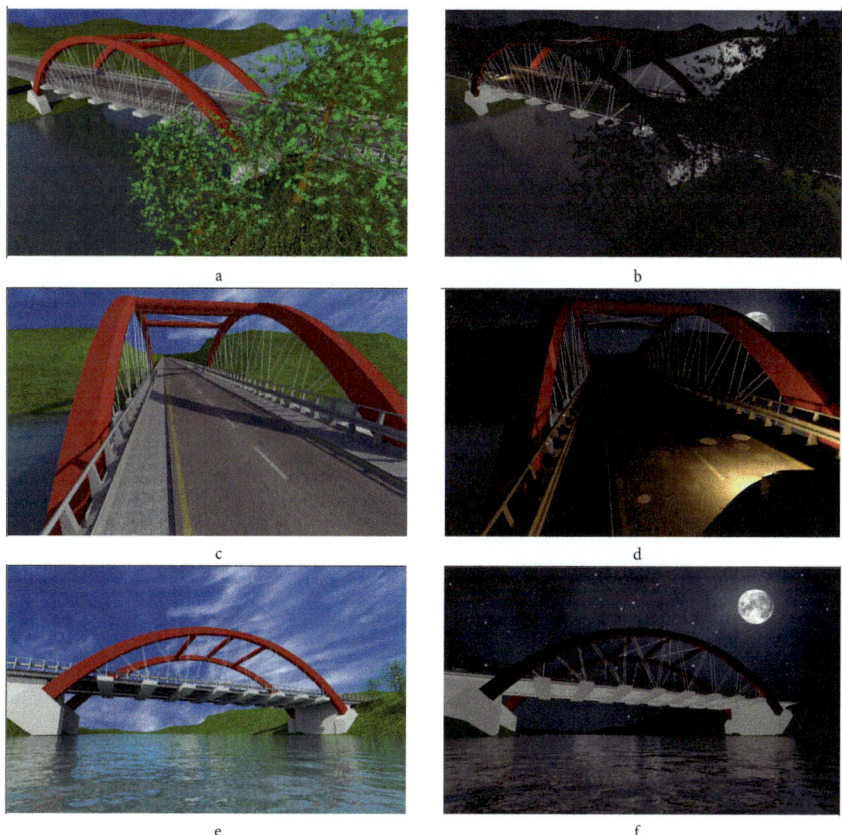

Fig. 4 Three takes of the bridge in the environment where it is to be accomplished, made on day-and-night mode

It is worth mentioning that, for designers and constructors, a digital model implemented in a computer program used for 3D modeling is a valuable source of information. The designed structure can undergo a detailed analysis of its operation, using a specific scope or using different types of materials, which, as such, turn the most optimal decisions easier to make.

5 Inspiration

But getting back to the very object—the bridge—which, basically, has one mission, that is to lead the road over an obstacle. It would seem that this sole requirement concerns only its structure, and that it is possible to achieve an unrestricted purity of form through nothing more than construction. So, why have we chosen bridges of string construction to achieve the didactic goal proposed? Bridges can be included in the group of modern buildings that spectacularly embody the support structure. This architectural example serves as a model for extracting plastic shapes, revealing the qualities of the materials and forces constraints on the exposed structure. Modern suspension and cable stable bridges seem to confirm Schopenhauer's thought that "purely architectural" means the same as "structural" [1].

More importantly, the form of string construction bridges that aims for the stability that is proper to a building, inspires the structural form which is seen as a particular kind of piece of art of designing and architectural composition. Construction elements included in the architectural order of things inspire the form of objects with string structure and determine relations with, and within, the environment.

6 Conclusion

Nowadays, we seem to witness some sort of rivalry in the execution of architectural objects. Newly established projects tend to be bigger and better, both structurally and technologically, than their predecessors. The use of computer methods stimulates the imagination and, consequently, acts as drive for creativity [2].

The results of the assignments given to students show that the Visualization of Building objects was effective in helping them gain a better understanding of spatial structures and to comprehend the relationship between structure and form in a more integrated way. From this exercise, we have concluded that such kind of work raises students' awareness about various structural concepts.

References

1. Sławińska, J.: Ekspresja sił w nowoczesnej architekturze. Arkady Press, Warsaw, Poland (1997)
2. Tofil, J., Pawlak-Jakubowska, A.: Architectural form and building material of suspension and cable-stayed bridges—visualization of geometrical structure. In: Scientific Proceedings of the 12th International Conference on Engineering Graphics Baltgraf 2013, Riga, Latvia, pp. 215–222 (2013)

Jolanta Tofil is doctor of technical sciences in architecture and urban planning discipline and is an assistant professor at the Faculty of Civil Engineering of the Silesian University of Technology. She specializes in research on architectural form of string construction: roofs and bridges as well as descriptive geometry and engineering drawings. She is author of publications in these fields. She is Secretary of the Polish Society for Geometry and Engineering Graphics.